Photochemical Smog and Ozone Reactions

Photochemical Smog and Ozone Reactions

Two symposia sponsored by
Divisions of the American
Chemical Society at the 161st
ACS Meeting at Los Angeles,
Calif., March 29, 1971 and
April 1, 1971.

ADVANCES IN CHEMISTRY SERIES **113**

AMERICAN CHEMICAL SOCIETY

WASHINGTON, D. C.   1972

ADCSAJ 113 1-285 (1972)

Library of Congress Catalog Card 72-96012

ISBN 8412-0155-2

PRINTED IN THE UNITED STATES OF AMERICA

Advances in Chemistry Series

Robert F. Gould, *Editor*

FOREWORD

ADVANCES IN CHEMISTRY SERIES was founded in 1949 by the American Chemical Society as an outlet for symposia and collections of data in special areas of topical interest that could not be accommodated in the Society's journals. It provides a medium for symposia that would otherwise be fragmented, their papers distributed among several journals or not published at all. Papers are refereed critically according to ACS editorial standards and receive the careful attention and processing characteristic of ACS publications. Papers published in ADVANCES IN CHEMISTRY SERIES are original contributions not published elsewhere in whole or major part and include reports of research as well as reviews since symposia may embrace both types of presentation.

CONTENTS

Photochemical Smog

A symposium sponsored by
the Division of Physical
Chemistry at the 161st
Meeting of the American
Chemical Society, Los Angeles,
Calif., March 29, 1971.

Bernard Weinstock
Symposium Chairman

Preface

In the early 1950's A. J. Haagen-Smit demonstrated that the irradiation of mixtures of nitrogen oxides and hydrocarbons produced products with properties similar to that of atmospheric photochemical smog. This pioneering effort stimulated much research on the problem. Improved techniques for studying photochemical reactions at concentration levels on the order of parts per million have been developed, and knowledge of the rate constants and mechanisms of the reactions involved has been advanced. New compounds have been discovered; however, a quantitative correspondence between laboratory studies and atmospheric observations has not been obtained. As a result, a reliable technical basis for the abatement requirements to alleviate photochemical smog in the Los Angeles Basin is still lacking.

On December 31, 1970, the Clean Air Amendents were passed into law by the Congress. The law expresses the urgency that Congress felt for the solution of the air pollution problem in this country. This sense of urgency is reflected in a number of unusual features of the law. For example, the technical feasibility of meeting the law had not and still has not been established. Also, an analytical model for the derivation of the abatement requirements seemed to be wanting. This can be inferred since the same factor of ten reduction was required for each of the pollutants without modification for the level of control already in force. Since hydrocarbons and carbon monoxide emissions had already been substantially reduced for 1970 motor vehicles, the net overall reduction factor from precontrol levels for these substances became 37 and 26, respectively. Perhaps, even more significant was the fact that the air quality standards that the law purported to achieve had not then been established and were not announced until April 1971.

Despite the important implications of the 1970 Clean Air Amendments and of the Air Quality Standards to the purpose of this symposium, for the most part the cogency of the program was unaffected. The five papers presented at this symposium that were selected for inclusion in this volume address themselves to significant aspects of the photochemical smog problem. Hopefully, this volume will stimulate a broader interest of the scientific community in the quantitative aspects of air pollution.

BERNARD WEINSTOCK

Dearborn, Mich.
November 7, 1972

Global Aspects of Photochemical Air Pollution

ELMER ROBINSON[a]

Stanford Research Institute, Menlo Park, Calif. 94025

The most serious effect of increasing photochemical air pollution on a global basis is the production of high concentrations of submicron aerosol in the atmosphere. This effect will first be noted in the midlatitude zone between 30° and 60° north and could cause large unfavorable changes in weather patterns and world climate. The probability that these changes may occur with unchecked increase in concentrations of photochemical pollutants sufficiently justifies comprehensive control of photochemical pollutant emissions. Since rigorous control programs have already been initiated in the United States and abroad, large increases in aerosol concentrations in the atmosphere rising from photochemical pollutant reactions are not likely, and the global problems related to photochemical air pollution are expected to remain overshadowed by the problems in affected local areas.

To consider the global aspects of photochemical smog risks extending the smog problems faced by Los Angeles until the dimensions are world-wide—until there would be no place to hide. Such a technological time-bomb is unlikely to exist if we do groundwork and establish perspectives of the present air pollution situation.

We are not in an unknown technological jungle; in 1944 photochemical smog was noticed in the Los Angeles Basin, existing as vegetation damage. Thus photochemical smog occurred at least 27 years ago. Also the Los Angeles County Air Pollution Control District is about 25 years old, having been formed because photochemical smog occurred in the Los Angeles Basin in the mid 1940's, and we know how to design and

[a] Present address: College of Engineering, Washington State University, Pullman, Wash. 99163.

operate control functions. Ozone, a critical component of photochemical smog, was identified in the Los Angeles Basin in the late 1940's, a significant first step for analyzing the photochemical smog reaction that was then occurring in the Los Angeles Basin. Also, this is approximately the 20th anniversary of Haagen-Smit's initial experiments in which photochemical reactions involving gasoline fumes were found to produce ozone and other adverse effects identified as photochemical smog. Thus, we have a situation which is slowly succumbing to technical investigation although there are still many mysteries left.

However, we are now concerned not just with Los Angeles smog but with the global aspects of photochemical smog. Smog has become a world-wide phenomenon with world-wide importance; discussions of the global aspects of photochemical smog are more than 15 years old. One of the first papers on this subject was titled "Global Aspects of Air Pollution as Checked by Damage to Vegetation" by Fritz Went (1). Here he recounted his investigations around numerous large cities throughout the world, and during these investigations he came to several important conclusions. Before 1955 Went had observed unmistakable smog damage to vegetation near the following cities: Los Angeles, San Francisco, New York, Philadelphia, Baltimore, London, Manchester, Cologne, Copenhagen, Paris, Sao Paulo, and Bogota. No damage had been seen in Houston, St. Louis, Amsterdam, Zurich, Madrid, Rome, or Jerusalem.

Photochemical smog damage was essentially local in nature. Until 1955 Went could find no smog-damaged plants in rural areas more than 50 miles away from major metropolitan areas, and only around Los Angeles and London did he observe plant damage as far away as 50 miles. In Paris damage to vegetation occurred as much as 12 miles from the center of the city. Resulting from his observations, Went set the approximate date when smog damage to vegetation became serious in the various metropolitan areas he visited. For example, in Los Angeles photochemical smog damage did not occur before 1944 while in the San Francisco Bay Area the first damage was noted in 1950. In Paris and New York the onset was 1952. Compiling his data on vegetation damage and comparing it with the use of gasoline in the various urban areas, Went concluded that when gasoline consumption exceeded 12 tons/square mile/day in a typical urban area, smog damage to vegetation probably occurred. Thus at least some aspects of global photochemical smog have been recognized and have been a matter of concern for many years; it is not some new crisis.

Actually a global smog problem is defined at least two ways. One concept was described in 1955 by Went—i.e., local centers of photochemical smog geographically distributed and related directly to major urban areas around the world—making the problem globally important.

If we tried to name all present locations where photochemical smog was a major concern or at least so considered by the population of the city, we would probably include almost all of the major urban areas in the world. Even places considered relatively remote—*e.g.*, Ankara, Turkey, and Seoul, Korea—have recently been publicized as major centers of serious smog problems; thus, photochemical smog is geographically a global problem.

The second definition of global air pollution is the geophysical one where the concern is with the impact of photochemical air pollution on the total atmospheric environment without specially considering any one urban area. This is analogous to our concern about how increasing carbon dioxide will affect the climate and the total atmospheric environment. Here we are really talking about the long term or geophysical importance of photochemical smog to our spaceship earth and the possibility of causing a serious perturbation in our total global environment; it should be closely scrutinized.

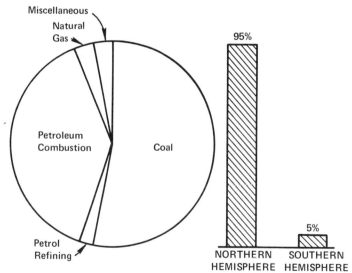

TOTAL NO$_2$ EMISSIONS = 53 x 10^6 T/yr

Figure 1.   Distribution of NO$_2$ pollutants

Pollution Sources

Nitrogen Oxide. The photochemical smog reaction involves nitrogen oxides, hydrocarbons, and sunlight. The global importance of this pollutant system depends upon the amounts of materials emitted to the atmosphere, their residence time in the atmosphere, and their reaction products.

Thus, we should examine the total emissions of nitrogen oxides and hydro-carbons, especially the so-called reactive ones, on a global scale and try to determine their impact on atmospheric conditions.

Combustion emissions are the major sources of nitrogen oxides in urban areas. The sources of NO and NO_2 are primarily those combustion processes where the temperatures are high enough to fix the nitrogen in the air and where the combustion gases are quenched rapidly enough to reduce the subsequent decomposition of NO. Figure 1 shows the esti-mated global distribution of nitrogen oxides pollution by source, expressed as NO_2. These estimates are based on fuel consumption and NO_2 produc-tion ratios. About 53×10^6 tons are estimated to be produced annually, with 51% of the total resulting from coal combustion and 41% resulting from petroleum production and the combustion of petroleum products. Within the petroleum class, combustion of gasoline and residual fuel oil is the major NO_2 source. In the coal combustion category power gen-eration and industrial uses account for most of the NO_2 emissions, ex-pected because of the high temperatures involved in most of these operations. Domestic and commercial use of coal are minor sources of NO_2. Natural gas combustion is also a relatively minor source of NO_2, only 4% of the total. Other miscellaneous sources, combustion of fuel wood and incineration, are also relatively insignificant. Figure 1 estimates how NO_2 pollutants are distributed between the northern and southern hemispheres. The 95–5% ratio between the northern and southern hemi-spheres is readily explained by the vast difference between combustion processes (2) and population densities in the two hemispheres.

On a global basis, however, these estimated pollutant emissions of NO_2 are probably a minor factor in the total circulation of nitrate com-pounds within the atmosphere. It is estimated (3) that natural emissions of NO_2 may be as much as 15 times greater than the pollutant emissions or more than 700×10^6 tons. This estimate of natural emissions is based on an estimated nitrogen cycle for the atmosphere, and it is believed to result from NO produced by biological reactions. Peterson and Junge (24) recently estimated natural nitrogen compound emissions may be no more than twice as large as currently estimated pollutant sources.

Hydrocarbons. Photochemical air pollution in many urban areas focuses attention on nitrogen oxides and hydrocarbons. Globally sources of hydrocarbons include not only the urban pollutant sources but also major natural sources.

Figure 2 shows one estimate of the global emissions of hydrocarbon pollutants. The total of 88×10^6 tons mainly represents petroleum-related operations—*i.e.*, gasoline usage, 34×10^6 tons, refinery operations, 6.3×10^6 tons, petroleum evaporation and transfer losses, 7.8×10^6 tons, and solvent usage, 10×10^6 tons—totaling 58×10^6 tons or 66% of the total. The estimate of solvent usage is rough, based on an estimate that world usage is three times that in the United States.

Incineration is probably a major contributor to organic emissions, assuming roughly that the world total is about five times that in the United States, and the U.S. incineration average is about 3 lb/person/day. Incineration is generally considered to be a poor combustion process and thus has a high calculated emission rate of 100 lb of hydrocarbons per ton of incinerated waste (4). This gives an estimated total emission for the United States of 25×10^6 tons. The combustion of wood as fuel however is estimated to be relatively efficient and a negligible source of hydrocarbon emissions; coal combustion gives about 3×10^6 tons of organic emissions, of which about two-thirds result from domestic and commercial combustion sources.

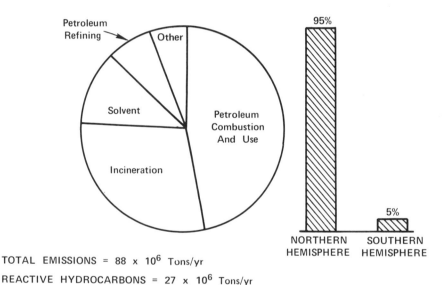

TOTAL EMISSIONS = 88 x 10^6 Tons/yr

REACTIVE HYDROCARBONS = 27 x 10^6 Tons/yr

Figure 2.   Distribution of hydrocarbon pollutants

In areas where photochemical air pollution is a serious problem, the olefins and other more reactive hydrocarbons are major concerns rather than just with the total organic emissions. Using factors mainly derived by the Bay Area Air Pollution Control District in San Francisco (5), the total emissions of organic materials have been broken down to give an estimate of the reactive hydrocarbon emissions. About one-third, or about 27×10^6 tons, are considered reactive out of the 88×10^6 total tons of organic materials; over half of the estimated reactive emissions result from automobile emissions.

There also are major sources of organic compounds from nature. Methane is a major emission from the natural environment and in non-urban atmospheres concentrations generally range between 1–1.5 ppm (6). The major source of atmospheric methane is the decomposition of organic material in swamps, marshes, and other bodies of water. Natural gas seepage possibly significantly contributes methane to the atmosphere in certain petroleum areas. It is estimated that natural sources of methane are about 1600×10^6 tons annually (7).

The biosphere is a major contributor to the atmosphere of heavier hydrocarbons. Fritz Went (8, 9), who first recognized the global extent of smog, pointed out the general importance of natural emissions from vegetation. He estimated that sources in the biosphere annually emit between 170×10^6 and 10^9 tons of hydrocarbon material to the atmosphere. Went also observed that these materials are mainly in the terpene class and that, because they are photochemically reactive, these materials are polymerized in atmospheric photochemical reactions to form an organic aerosol. He attributes the blue haze found in many forested areas to the optical effects of this aerosol.

Aerosol Production and Distribution

In the short term sense, reactions involving photochemical smog that attract the most attention are the nuisance factors producing eye irritation and vegetation damage. However, on a global scale reactions involving more persistent products are primarily important—*i.e.*, the submicron aerosol particles. A significant part of the chemical scavenging from the atmosphere occurs after reactions have formed aerosol particles. The forming of an aerosol is believed to be a principal step in scavenging the components of photochemical smog reactions and also for sulfur and nitrogen compounds. Initially aerosols formed by these reactions are small particles; however, significant coagulation to form larger particles up to about 0.1 micron radius occurs because of Brownian motion. These larger particles are then incorporated into cloud, fog, and raindrops and eventually are removed from the atmosphere by precipitation. The significant chemical content of rain fall is evidence of the important scavenging role of precipitation. Particles are removed from the atmosphere by dry deposition through gravitational settling.

Although it has been recognized for many years that smog reactions form important amounts of aerosol particles, little is known about the actual reaction mechanisms that form these aerosols. It is postulated that the reactive materials from combustion sources and from the biosphere—*e.g.*, terpenes—undergo similar rapid chemical changes in the presence of nitrogen oxides and sunlight, the familiar photochemical smog

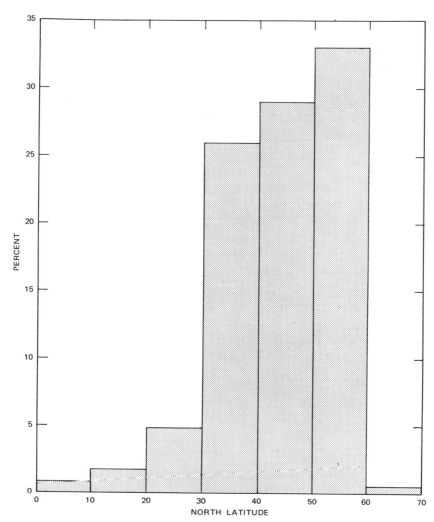

*Figure 3.   Northern Hemisphere latitudinal distribution of total energy con-
sumption (percent of 1968 global energy consumption)*

reaction. In a series of free radical polymeric reactions, these reactive
organics are transformed into condensed aerosol particles in the atmosphere. We estimate that the mass of aerosols formed by terpene reactions are generally equivalent to the mass of the terpenes or reactive
organics in the initial gas phase. Thus, globally, we predict that the
photochemical aerosols will amount to about 30×10^6 tons annually from
pollutant sources and about 200×10^6 tons of aerosol material as a result
of the natural emissions of terpene-type organic materials to the atmosphere.

It is our estimate that all pollutant sources result in about 300×10^6 tons per year of particulate material being introduced into the atmosphere and that pollutant and natural sources combined amount to about 3000×10^6 tons per year. Thus, on a global scale 30×10^6 tons per year of pollutant photochemical aerosols are about 10% of the total pollutant aerosols and about 1% of the total annual atmospheric aerosol production. The 200×10^6 tons of aerosol particles resulting from the photochemical scavenging reactions involving natural organic emissions is somewhat less than 10% of the total aerosols emitted to the global atmosphere.

Here aerosol concentrations have been referred to as general global average values even though the actual world-wide situation is one of great heterogeneity regarding pollutant and natural particle sources. Natural and pollutant sources are not uniformly distributed, and thus the particles produced have a greater relative impact on the environment of one geographical area than on another.

Pollutant emissions on a country-by-country basis are not readily available; however, Figure 3 shows the 1968 consumption of energy in the northern hemisphere classified into 10° latitude bands in units of millions of tons of coal equivalent energy (*10*). Since energy consumption and pollutant emissions, aerosols and gases, are assumed to be reasonably well correlated, atmospheric aerosols from particulate emissions and scavenging reactions should also show a latitudinal distribution that is similar to Figure 3. In this tabulation where countries such as the United States covered more than one 10° band, the energy consumption was proportionately divided according to the area in each band. Energy consumption data indicate that 96% of the global consumption of energy occurs in the northern hemisphere and 86% in the latitude band between 30°–60° north. This imbalance in latitudinal distribution of energy consumption is shown in Figure 3. For the United States assumption of a correlation between energy and pollutant particle production checks reasonably well; in 1968 the United States consumed 34% of the energy consumed in the northern hemisphere and produced about 31% of the particulate materials.

Particle emissions resulting from natural processes should be roughly distributed according to the relative amounts of land and ocean areas in various latitudinal bands. Assuming this, natural amissions are estimated to be a function of latitude, based on the amount of land and water distributed in various latitude zones. When this is done for the northern hemisphere and compared with an estimated zonal distribution of pollutants based on Figure 3, the relative contribution of pollutant sources to atmospheric aerosol concentrations as a function of latitude is estimated.

Figure 4 shows the results of such a calculation. These data indicate that pollutant aerosol sources are probably important on a global scale only within the latitude band from 30°–60° north. In this 30° zone, the average pollutant contribution is about 31% of the total atmospheric aerosol concentration. Since photochemical smog reactions account for about 10% of the total pollutant aerosol generated, photochemical smog is estimated to be responsible for about 3% of the pollutant contributions of aerosols within the 30° band from 30°–60° north. Throughout the northern hemisphere pollutant sources account for about 17% of the total atmospheric aerosol. Using the 10% figure as that part assigned to photochemical aerosols, we get a value of about 2% of the total northern hemisphere aerosol to result from photochemical smog reactions. In the southern hemisphere, the contribution from all pollutant sources is only about 1% of the total atmospheric aerosol produced.

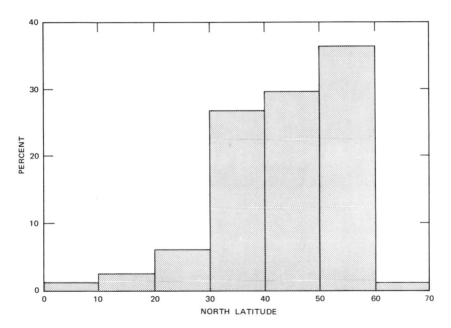

Figure 4.   Relative contribution of pollutant aerosols to total atmospheric aerosol concentrations in the Northern Hemisphere

These estimated latitudinal distributions are modified somewhat by winds carrying pollutants downwind from the source areas. However, the general west-to-east movement of air masses and the concentration of pollutants into the lower portion of the troposphere tend to minimize wind trajectory effects on the patterns shown in Figures 3 and 4.

Geophysical Effects

Although photochemical pollutant aerosols apparently account for only a few percent of the total atmospheric particles, the geophysical importance of pollutant aerosols is still important and involves two questions.

1. To what extent are current emissions of pollutants causing measurable increases or accumulations of aerosol particles in the atmosphere?

2. How may presently observed or future change in atmospheric aerosol concentrations affect the earth's climate?

If the previous figure showing the maximum impact of pollutant emissions in the zone between 30°–60° north is correct, we expect to find persistent downwind pollutants indicated first in this latitude zone over the North Atlantic where the effects of the concentrated pollutant sources in North America are most likely seen. Pollutant aerosols are found as shown by observing that fly ash particles have been collected across the North Atlantic (*11*). These apparently came from sources on the east coast of the United States and Canada. These fly ash particles were relatively large, greater than 2 microns, and concentrations were extremely low, but the indications are that this material must have been carried for considerable distances over the Atlantic and accompanied by other less easily identified pollutants.

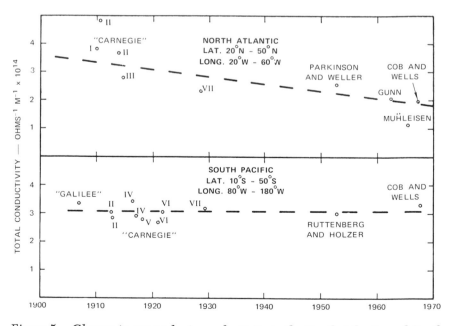

Figure 5.   Change in atmospheric conductivity in the North Atlantic and South Pacific between 1907 and 1967

Electrical conductivity data taken over the Atlantic over a period of years apparently show the impact of North American pollutant sources of fine particles on the Atlantic atmosphere (*12*). Atmospheric conductivity measurements can be related to the concentration of fine particles; those in the Aitken nuclei range from about 2×10^{-7} to 2×10^{-4} cm because particles in this size range act as a sink to remove from the atmosphere the small ions which determine the electrical conductivity. The greater the concentration of Aitken nuclei, the lower the conductivity because the concentration of small ions is reduced. Cobb and Wells (*12*) have presented data on electrical conductivity in the North Atlantic and in the South Pacific covering the period from about 1907–1967. Their results are shown in Figure 5 and indicate that the conductivity over the North Atlantic seems to have decreased by at least 20% in the past 60 years and by about 15% since about 1929. According to these authors, this change in conductivity might be equivalent to doubling the fine particle aerosol concentration over the North Atlantic area. The data from the Pacific Ocean area in the southern hemisphere indicate that in this region aerosol concentrations in the 1960's are no higher than they were in the period from about 1910–1929. Gunn (*13*) came to similar conclusions, increasing concentrations of aerosol material over the Atlantic were presumably coming from increased pollutant sources in North America, and aerosol concentrations were generally stable over the South Pacific. While aerosol concentrations may have increased over the North Atlantic, similar trends have not persisted over more remote ocean areas in the southern hemisphere, and thus it is concluded that truly global increases of atmospheric particle concentrations have probably not occurred. Similar data for the North Pacific would be very valuable but apparently are not available.

Apparent changes in atmospheric fine particle concentrations within the 30° to 60° latitude zone in the northern hemisphere have been pointed out by McCormick and Ludwig (*14*) and by Flowers *et al.* (*15*), based on turbidity measurements made in the United States and Europe. The absence of detectable turbidity changes over a period of 15 years in the southern hemisphere is shown by data presented by Fischer (*16*). These two observations confirm our earlier remarks relative to the separating of the northern and southern hemispheres and that aerosol effects are noticed most in the mid-latitudes of the northern hemisphere.

Turbidity data applied to upper atmospheric conditions, have been gathered at the meteorological observatory on Mauna Loa, Hawaii and presented by Peterson and Bryson (*17*). These data covered about a 10-year period from 1957–1967 and have been interpreted by Peterson and Bryson to indicate a general increase in the fine particle concentrations, as well as showing the effects of one or more major volcanic eruptions. Mitchell (18), however, attributes the change in these observed turbidity values from Hawaii to accumulated volcanic activity affecting stratospheric aerosol concentrations rather than to the impact of pollutant aerosols.

Whereas photochemical aerosols are expected to account for roughly 10% of the total pollutant aerosol load, several factors related to the total aerosol distributions combine to make significant global increases of pollutant aerosol concentrations difficult to attain. Probably the most important is that almost all of the particulate material in the atmosphere results from surface sources and subsequently is mostly found in the lowest few kilometers of the atmosphere. At this low altitude particles are exposed to repeated condensation and precipitation activity which acts as a relatively effective scavenger of aerosol material. An average residence time of three days is reasonable for this situation and has been used in our model calculations (19). In this period of time, the mean winds in the zone 30° to 60° north, as described by Mintz (20), would carry the aerosol cloud between 1,000–2,000 miles from west to east. This would be far enough to cause the effects noted by Cobb and Wells over the North Atlantic, but on a global scale the result would be negligible. The general lack of atmospheric transport mechanisms to move tropospheric air masses across the equator is important in preserving the isolation of the two hemispheres.

Aerosol scavenging processes are primarily accomplished by precipitation, and as such the process is one of continual dilution. Thus although most of the particles may be removed within a relatively few days, there is some residual material which persists in the atmosphere for much longer times. The ability to detect such material depends upon whether the specific particulate material can be distinguished from the natural background. Lead aerosols from pollutant sources are detected after extensive dilution because of the negligibly small background of lead that exists in the atmosphere. This is shown by measurements over the central northern Pacific (21) of lead aerosol concentrations of the order of 0.001 $\mu g/m^3$, which is a dilution of 3–4 orders of magnitude compared with pollutant urban air mass concentrations of 1–10 $\mu g/m^3$. Chow argues that this concentration represents the residual concentration of pollutant lead aerosols in the northern hemisphere. Compared with a typical remote-area dust loading of 10 $\mu g/m^3$, this lead concentration represents about 0.01% of the total particulate material contained in a remote atmosphere. If as a first approximation photochemical aerosols in urban atmospheres are roughly comparable in size and concentration with lead aerosols, we estimate that in remote areas the photochemical aerosol contribution from pollutant sources amounts to a possible change of

0.01% of the total particulate loading in the atmosphere, an undetectable amount.

Besides problems of local annoyance, a major concern about atmospheric pollutants is whether or not their presence in the atmosphere could cause changes in the earth's climate. The observed increases in carbon dioxide content in the atmosphere first raised this concern because, at least in theory, increased carbon dioxide concentrations could cause increased absorption of infrared radiation and could increase the temperature of the atmosphere. If significant warming were to result, changes in climate on a global scale could follow. For changes in the fine particle concentration in the atmosphere, the simplified argument as described by McCormick and Ludwig (14) is that increased particle concentrations could lower global temperatures because solar energy would be reflected back to space before it reached the earth's surface. Charlson and Pilat (22) and others have questioned this simplified approach and show that aerosols in the lower atmosphere can cause warming. Evidence from large scale phenomena—*e.g.*, such as volcanic eruptions—is that high concentrations of particulate material in the stratosphere can reduce temperatures at ground level (23). The physical characteristics of the aerosol particles are also important in whether or not warming or cooling is caused.

Our understanding of the atmospheric climate is not sufficient for us to model in detail the effects of such conditions as atmospheric aerosols or CO_2 concentrations, and we must be satisfied with general models of these conditions. Currently it has not been possible to include the fact that particulate pollutants are not evenly distributed on a global basis but are primarily confined to a single relatively narrow latitude zone, as pointed out in Figures 3 and 4. We conclude that it is unlikely that there has been any climatic effect resulting from the concentrations of pollutant aerosols that have occurred within this zone in the atmosphere. We also conclude that any effects that are likely to occur in the future will result from the effects of enhanced differences between air masses rather than from a uniform change over global or hemispheric areas. Thus increased aerosol concentrations in specific zones may bring about increased temperature differences and shifts in storm tracks and precipitation patterns. Such a situation could result in a climatic change, but certainly not a simple one. However, even though we are generally unsure as to what the result might be or even the direction of possible change, if there were a long term and significant increase in the pollutant content of the atmosphere either of particles or of carbon dioxide, the potential damage to the global environment could be severe. Even the remote possibility of such an occurrence justifies concern about the long term or global impact of pollutants on the atmosphere.

Summary

Photochemical air pollution has a global impact in at least two ways:

1. Major urban areas around the world are increasingly becoming damaged and annoyed by photochemical smog. (This is a simple multiplication of local smog problems.)

2. Photochemical aerosols, only about 10% of the total pollutant aerosol material, can contribute to long distance, downwind effects from pollutant sources, and as part of the total pollutant aerosols probably have contributed to changes in the aerosol background concentration over the North Atlantic.

If these atmospheric impacts are not remedied and some controls established over photochemical aerosols as well as other pollutant sources, gradually increasing concentrations will be observed in the mid-latitude zone between 30° and 60° north and then gradually outward throughout much of the northern hemisphere. If such a change should occur and aerosol concentrations continue to increase, the impact on the global atmosphere could be severe, bringing about changes in weather patterns and subsequent changes in climate. Although the direction of such change is unpredictable at this time, the probability is that changes of this type could be severe, and it justifies controlling emissions that lead to aerosol formation.

The probability of pollutant aerosol concentrations becoming concentrated enough to be a major climatic hazard seems to be small because such a change would depend on our permitting urban area conditions to deteriorate probably by an order of magnitude. Considering the present concern of the general public in the United States and abroad for the atmospheric environment, such an increase seems unlikely. Also, regulatory programs are now being implemented that should reduce the quantities of photochemical and non-photochemical aerosols. Thus, the global aspects of photochemical smog seem to be much less critical than are the problems in affected local areas.

Acknowledgment

This discussion was mainly drawn from research on the global distributions of gaseous and particulate air pollutants sponsored at Stanford Research Institute by the American Petroleum Institute.

Literature Cited

1. Went, F., *Proc., Nat. Air Pollution Symp., 3rd,* Stanford Research Institute, Pasadena (1955).
2. United Nations, "Statistical Papers," *Series J, No. 11,* World Energy Supplies, 1963–1966, U.N., New York (1968).
3. Robinson, E., Robbins, R. C., *J. Air Poll. Cont. Assoc.* (1970) **20**, 303.

4. Duprey, R. L., "Compilation of Air Pollutant Emission Factors," Environmental Health Series, Air Pollution, 1968.
5. Bay Area Air Pollution Control District, San Francisco, unpublished emission estimates, 1966.
6. Junge, C. E., *Air Chemistry and Radioactivity*, Academic, New York, 1963.
7. Robinson, E., Robbins, R. C., "Sources, Abundance, and Fate of Gaseous Atmospheric Pollutants," American Petroleum Institute, 1968 and 1969.
8. Went, F. W., *Proc. Nat. Acad. Sci.* (1960) **46,** 212.
9. Went, F. W., *Tellus* (1966) **18** (203), 549–556.
10. United Nations, Statistical Papers," *Series J, No. 11,* World Energy Supplies 1963–1966, UN, New York (1968).
11. Parkin, D. W., Phillips, D. R., Sullivan, R. A. L., *J. Geophys. Res.* (1970) **75,** 1782–93.
12. Cobb, W. E., Wells, H. J., *J. Atmos. Sci.* (1970) **27,** 814–19.
13. Gunn, R., *J. Atmos. Sci.* (1964) **21,** 168.
14. McCormick, R. A., Ludwig, J. H., *Science* (1967) **156,** 1358.
15. Flowers, E. C., McCormick, R. A., Kurfes, K. R., *J. Appl. Meteorol.* (1969) **8,** 955–62.
16. Fischer, W. H., Lodge, J. P., Pate, J. B., Cadle, R. D., *Science* (1969) **164,** 66.
17. Peterson, J. T., Bryson, R. A., *Science* (1968) **162,** 120.
18. Mitchell, J. M., Jr., "Global Effects of Environmental Pollution," p. 139, S. F. Singer, Ed., Reidel, Dordrecht-Holland, 1970.
19. Martell, E. A., *Advan. Chem. Ser.* (1970) **93,** 138.
20. Mintz, Y., *Bull. Amer. Meteorol. Soc.* (1954) **35,** 208.
21. Chow, T. J., Earl, J. L., Bennett, C. F., *Environ. Sci. Technol.* (1969) **3,** 737.
22. Charlson, R. J., Pilat, M. J., *J. Appl. Meteorol.* (1969) **8,** 1001–2.
23. Humphreys, W. J., "Physics of the Air," 3rd ed., Part V, Chap. III and IV, McGraw-Hill, New York, 1940.
24. Peterson, J. T., Junge, C. E., "Mom's Impact on the Climate," pp. 310–320, W. H. Matthews, W. W. Kellogg, and G. D. Robinson, Eds., MIT Press, Cambridge, 1971.

RECEIVED April 28, 1971.

2

Mechanisms of Smog Reactions

H. NIKI, E. E. DABY, and B. WEINSTOCK

Scientific Research Staff, Ford Motor Co., Dearborn, Mich. 48121

A model is developed to account for the chemical features of photochemical smog observed in laboratory and atmospheric studies. A detailed mechanism consisting of some 60 reactions is proposed for a prototype smog system, the photooxidation in air of propylene in the presence of oxides of nitrogen at low concentrations. The rate equations for this detailed mechanism have been numerically integrated to calculate the time–concentration behavior of all the constituents of the system. The model has been used to examine the effects of varying relative and absolute concentrations of the reactants. The conclusions of this examination provide a framework for the analysis of the more complicated atmospheric problem. Some of the key questions related to the atmospheric chemistry have been discussed in terms of the detailed model.

In recent years, a number of reaction models have been proposed to account for the chemical features of photochemical smog observed in atmospheric and laboratory studies (*1, 2, 3, 4, 5, 6, 7, 8, 9, 10, 11*). Because of the complexity of smog chemistry and a lack of detailed knowledge of many relevant elementary reactions, numerous assumptions and simplifications are made in these mechanistic interpretations. A model for the chemistry of smog is presented here with a critical evaluation of the factors that control the major course of the reactions. The photooxidation of propylene (C_3H_6) in the presence of nitric oxide and nitrogen dioxide ($NO + NO_2 = NO_x$) is used as a prototype for this study.

Propylene is a good choice for this purpose because it is the simplest hydrocarbon that displays the major characteristics of photochemical smog, and much experimental smog chamber data on its photooxidation in the presence of NO_x have been reported (*12, 13, 14, 15, 16, 17, 18,*

19, 20, 21, 22, 23, 24). However, even for this simple smog system, over 150 possible elementary reactions have been examined, and of these some 60 important reactions have been selected for the present model. Since the relevant elementary reactions have been previously reviewed (*25, 26, 27, 28, 29*), only the essential features af this kinetic model will be discussed here. These 60 reaction steps have been numerically integrated by computer to evaluate the concentration–time behavior of the system with respect to a number of critical parameters. At first, the numerous unknown rate constants and mechanisms seem to provide sufficient degrees of freedom to permit any arbitrary parametric fit to the experimental data. However, here the arbitrariness is largely removed by detailed consideration of the complex concentration–time behavior of reactants and products. This imposes several stringent boundary conditions on the flexibility of the model. Furthermore, we will prove that relatively few elementary reactions are crucial in determining the major characteristics of the smog reactions. Some of the crucial parameters that pertain to the $C_3H_6-NO_x$ system and to other relevant smog systems will be discussed below. All the elementary reactions included in the model are listed in Table AI. The photochemical roles of chlorine, sulfur oxides, metastable oxygen molecules, and aerosols (*26, 27*) are matters of current interest but are not considered here.

Kinetic Features of Smog Chamber Data: $C_3H_6-NO_x-Air$ System

The data shown in Figure 1 were used as a reference for the kinetic scheme formulated in this work (*12*). In this experiment a mixture of 2.23 ppm C_3H_6, 0.97 ppm NO, and 0.05 ppm NO_2 in prepurified air (50% relative humidity) was irradiated at 31.5 ± 2°C by simulated sunlight with an intensity corresponding to a rate of NO_2 photodissociation in N_2, k_d, of 0.4 min^{-1}. The reactant concentrations used in this experiment are typical of smog chamber studies but are an order of magnitude higher than atmospheric levels. The implications of this will be discussed below.

The general kinetic features of the concentration–time behavior shown in Figure 1 are summarized as follows: in the initial stage the reaction rates of C_3H_6 and NO are markedly slower than in the succeeding stages of reaction, which suggests an induction period. The NO is then rapidly oxidized to NO_2. The concentration of NO reaches a low, steady level which persists in the later stages. Measurable ozone (O_3) formation starts about the time that the NO_2 concentration reaches a maximum. The subsequent loss of NO_2 is partially accounted for by the formation of peroxyacetylnitrate (PAN). Small amounts of other organic nitrates and nitrites have been reported for similar systems, but not enough to account for the rest of the NO_2 consumed (*15, 16, 17, 18, 19, 20*). Presumably the missing NO_2 is largely converted to nitric acid (HNO_3) (*30*). Formaldehyde ($HCHO$) and acetaldehyde (CH_3CHO) are formed in nearly stoichiometric amounts as the major oxidation products of C_3H_6 although only the acetaldehyde data are shown in Figure 1. The remaining carbon containing products are mainly CO and CO_2.

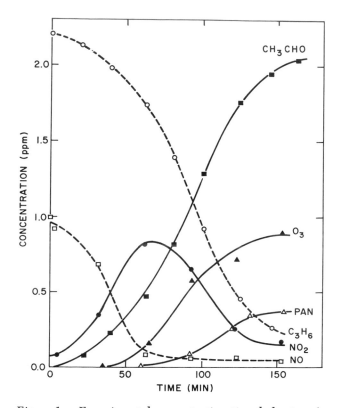

Figure 1. Experimental concentration–time behavior of the reactants and some of the products in the photo-oxidation of C_3H_6 in the presence of NO and NO_2 in air, obtained by Altshuller et al. (12, 13).

It should be emphasized that smog chamber data are susceptible to a number of experimental uncertainties. Analytical techniques, impurity effects, and wall effects are major sources of error which are often difficult to determine quantitatively, and the resulting uncertainties must be considered in the derivation of a kinetic model. For the C_3H_6–NO_x system, there is a fair degree of consistency among available smog chamber data (*12, 13, 14, 15, 16, 17, 18, 19, 20, 21, 22, 23, 24*). Nevertheless, further improvement of the methodology of smog chamber experiments is desirable for detailed kinetic analysis. The complexity of the concentration–time dependence requires that data be obtained for a larger variety of initial conditions than has been reported. Part of the difficulty arises because the large mass of data precludes publishing detailed profiles. These data cannot be adequately described by meaningful, simple rate expressions, and as a result the customarily reported parameters—*e.g.*, reactant half-lives, rate of production, and terminal concentrations—are of limited value for detailed mechanistic interpretation. The published data of Altshuller *et al.* (*13*) are also incomplete in many of the aspects mentioned above; however, Altshuller supplied details that were used in this analysis (*12*). This particular set of data then represents a minimal set from which a detailed kinetic analysis is possible. It would be desirable to vary the reaction parameters and obtain data in similar detail before this study is regarded as complete. This study has not relied only on Altshuller's experiment; also, other published experimental data have been used and are discussed with appropriate references.

Introductory Kinetic Analysis

The final kinetic scheme describing the propylene–NO_x photooxidation is quite complex, and the qualitative, determining characteristics of the system become obscured in the computational detail. A simplified kinetic scheme is given below that provides a framework for the overall description.

$$NO_2 + h\nu \rightarrow NO + O \qquad k_1 = 0.4\ \text{min}^{-1} \tag{1}$$

$$O + O_2 + M \rightarrow O_3 + M(=\text{air}) \quad k_2 = 1.96 \times 10^{-5}\ \text{ppm}^{-2}\,\text{min}^{-1} \tag{2}$$

$$O_3 + NO \rightarrow NO_2 + O_2 \qquad k_3 = 29.3\ \text{ppm}^{-1}\,\text{min}^{-1} \tag{3}$$

$$O + C_3H_6 \rightarrow \text{stable products} \quad k_0 = 4.5 \times 10^3\ \text{ppm}^{-1}\,\text{min}^{-1} \tag{4}$$

$$O_3 + C_3H_6 \rightarrow \text{stable products} \quad k_{03} = 1.8 \times 10^{-2}\ \text{ppm}^{-1}\,\text{min}^{-1} \tag{13}$$

Rate constants and references are given in Table AI. The unit of parts per million (ppm) at atmospheric pressure and 300°K equals 2.45×10^{13} molecule cm^{-3}. For this introductory analysis k_0 has been taken as the total rate of reaction of O with C_3H_6 ($k_0 = k_{4a} + k_{4b} + k_{4c}$) and k_{03} the total rate of C_3H_6 ozonolysis ($k_{03} = k_{13a} + k_{13b}$).

This mechanism is a simplified scheme partially because the oxidation products of C_3H_6 by O and O_3 are assumed to be unreactive while actually they are important in the overall reaction scheme. The above equations, however, outline the primary steps that follow directly from the NO_2 photolysis, which is the driving force of the whole system (25). Several important kinetic relations are then derived from these reactions which are generally applicable to smog conditions. First, photostationary states of O and O_3 are expressed by Equations I and II. Square brackets designate concentrations.

$$[O] = \frac{k_1 [NO_2]}{k_2[O_2][M] + k_0 [C_3H_6]}, \tag{I}$$

and

$$[O_3] = \frac{k_2[O_2][M][O]}{k_3[NO] + k_{0_3} [C_3H_6]}. \tag{II}$$

Since $k_2[O_2][M] >> k_0 [C_3H_6]$ in all cases of interest, Equation 1 is further simplified to

$$[O] = \frac{k_1[NO_2]}{k_2[O_2][M]} \tag{III}$$

For Equation II a similar simplification is possible when $k_3[NO] >> k_{0_3}[C_3H_6]$

$$[O_3] = \frac{k_1[NO_2]}{k_3[NO]} \tag{IV}$$

This equation generally applies except in the later stages of the photochemistry in the laboratory and in the atmosphere. For example, in Figure 1 $[C_3H_6]/[NO]$ approaches 100, which would result in a 10% error by using Equation IV instead of Equation II.

According to Equations III and IV, [O] is determined by the light intensity and $[NO_2]$ whereas $[O_3]$ is a function of the light intensity and the ratio, $[NO_2]/[NO]$. For the light intensity of the experiment of Altshuller et al. (13), the ratio, k_1/k_3, is about 10^{-2} ppm. The validity of Equation IV has been shown by Stedman et al. (21). The validity of Equation IV under atmospheric conditions has been questioned (1), but the deviations may result from experimental artifact. In the data of Figure 1 when $[O_3]$ is the order of 0.1 ppm and $[NO_2]/[NO]$ is ten, the agreement is quite good. However, when [O_3] is the order of 1 ppm, [NO_2]/[NO] becomes 100, and the experimental uncertainty for [NO] is too great to make a meaningful evaluation. However, with the use of the improved NO–O_3 chemiluminescence detector (*31*), Daby *et al.* (*75*) were also able to measure that high ratio with improved accuracy.

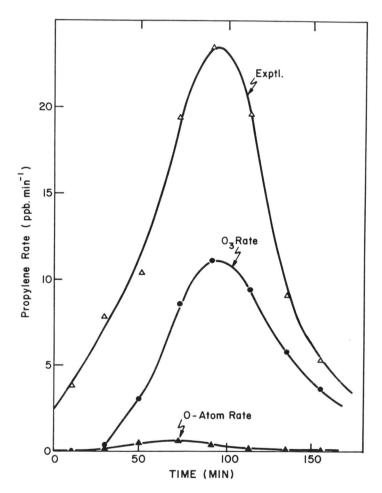

Figure 2. Comparison of the experimentally observed rates of C_3H_6 consumption with calculated rates of C_3H_6 reaction with O-atom and O_3.

The experimental rates are derived from the data of Figure 1. The O-atom rates are calculated from the differential rate equation for Reaction 4 and Equation III, using the experimental concentrations of C_3H_6 and NO_2. The O_3 rates are calculated from the differential rate equation for Reaction 13, using experimental concentrations of C_3H_6 and O_3.

The ratio of the reaction rate of O_3 with C_3H_6 (R_{O_3}) to that of O with C_3H_6 (R_0) is estimated from this reaction scheme to be:

$$\frac{R_{O_3}}{R_0} = \frac{k_2 k_{O_3} [O_2] [M]}{k_3 k_0 [NO]} \qquad (V)$$

which is equal to $0.54/[NO]$. In Figure 1 the rate of the O atom reaction with propylene is initially more important than that of O_3 when [NO] is 1 ppm but becomes less important rapidly as the reaction proceeds. This effect of the concentration of NO is a point of departure between smog chamber studies and atmospheric behavior because the [NO] is generally much lower than 1 ppm under atmospheric conditions.

A quantitative comparison of the observed rate of C_3H_6 removal (Figure 1) with rates calculated for O and O_3 reactions in this simplified scheme is shown in Figure 2. Several kinetic features are important in this figure. The O-atom rate is generally a small fraction of the O_3-rate, except at the beginning of the reaction when both rates are small. The combined $O + O_3$ rate is always less than the experimentally observed rate. The ratio of the observed rate to that of the combined $O + O_3$ rates is large during the early stage of the reaction. When [NO_2] is a maximum, this ratio is about a factor of two and approaches unity in the final stages of the reaction. These observations are consistent with the presence of a chain mechanism that dominates the early stages of the C_3H_6 oxidation but which becomes less important in the later stages when the observed rate largely results from ozonolysis.

This difference between the observed rate and the $O + O_3$ rates has been called the excess rate (25). The preceding discussion showed that a simplified model cannot explain this difference. Also the simplified model cannot explain the rapid conversion of NO to NO_2 that is characteristic of this system. A detailed mechanism that quantitatively accounts for the experimental data is presented below.

Detailed Mechanism

A detailed reaction scheme has been formulated to account for the laboratory data of Altshuller *et al.* (*12*). In Figure 3 the calculated concentration–time behavior based on this scheme is compared with the experimental data for C_3H_6, NO, NO_2, and O_3. More than 150 elementary reactions were initially considered in constructing this model. About 50 of these were eliminated from the scheme because their rate constant and/or the relative concentrations of the reactants are too small to be competitive with other reactions. Another 20 reactions were discarded

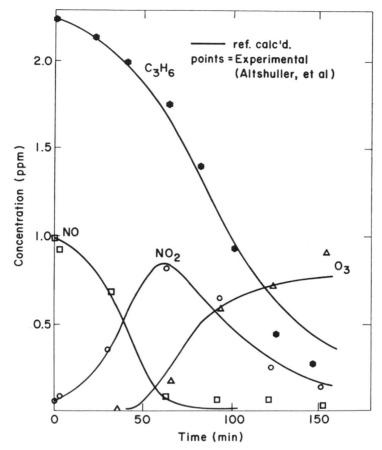

*Figure 3. Comparison of the concentration–time behaviors of
C_3H_6, NO, NO_2, and O_3 calculated from the model with the experimental data of Figure 1.*

The lines are calculated; the points are experimental.

because the resulting reaction products have not been observed experimentally. Of the remaining reactions 60 were selected, based on numerical analysis of the data of Altshuller *et al.* as well as that of other
laboratories (*15, 16, 17, 18, 19, 20, 21, 22, 23, 24*). The rate constants
and mechanisms are not well established for all of these 60 reactions.
However, as a result of over 100 computational tests, it has been established that only about a dozen of these are essential to derive the main
features of the C_3H_6–NO_x system. The role of these reactions and other
experimental parameters are quantitatively analyzed below.

Oxidation of Propylene by Atomic Oxygen. It is well known that
$O(^3P)$ atoms add to olefinic hydrocarbons to form an adduct, but the

subsequent fate of this adduct under atmospheric conditions is still uncertain (*26, 32, 33, 34, 35, 36, 37, 38, 39, 40, 41, 42, 43, 44*). Presumably the initial biradical adduct rapidly rearranges itself to an excited molecule which then either collisionally deactivates to form stable products or fragments to form radicals. The ratio of deactivation to decomposition greatly depends on the olefin and on the pressure and nature of the diluent gas. For propylene the yield of the fragmentation products, CHO and C_2H_5, is about 30% in N_2 at atmospheric pressure (*33*). A further interesting possibility exists when O_2 is used as the diluent gas. The O_2 might undergo chemical reaction with the excited molecules, yielding products similar to those of ozonolysis (*34*). This possibility of pseudo-ozonolysis could be particularly important for internally bonded olefins. These internally bonded olefins are known to be reactive in smog, and part of their reactivity could be from pseudo-ozonolysis because their O-atom adducts are almost always stabilized at atmospheric pressures and cannot otherwise initiate chain reactions. However, definitive evidence for this possibility is still lacking, and in the present kinetic scheme for the propylene oxidation by O-atoms, the role of O_2 was assumed to be identical to that of N_2.

$$O + CH_2{=}CHCH_3 \rightarrow \left[\begin{array}{c} O\cdot \\ | \\ \overset{\cdot}{CH_2CHCH_3} \end{array} \right]^* \begin{array}{l} \rightarrow CH_3CH_2\cdot + H\overset{\displaystyle O}{\overset{\|}{C}}\cdot \quad (4a) \\[2ex] \overset{M}{\rightarrow} CH_3CH_2\overset{\displaystyle O}{\overset{\|}{C}}H \quad (4b) \\[2ex] \overset{M}{\rightarrow} CH_3\overset{\displaystyle O}{\overset{\diagup\diagdown}{CH{-}CH_2}} \quad (4c) \end{array}$$

The stabilized products, propionaldehyde and propylene oxide, have not yet been observed in smog chamber experiments of this system. However, this is understandable because as has been discussed above, Reaction 4 accounts for only a small fraction of the total C_3H_6 consumption.

The fragment radicals, CHO and C_2H_5, undergo a number of radical transfer reactions involving O_2 and NO as given below:

$$CH_3CH_2\cdot + O_2 \quad = CH_3CH_2OO\cdot \quad (5)$$

$$CH_3CH_2OO\cdot + NO = NO_2 + CH_3CH_2O\cdot \quad (6)$$

$$2CH_3CH_2OO\cdot \quad = 2CH_3CH_2O\cdot + O_2 \quad (7)$$

$$CH_3CH_2O \cdot + O_2 = CH_3\overset{\overset{O}{\|}}{C}H + HO_2 \cdot \qquad (9)$$

$$HO_2 \cdot + NO = NO_2 + \cdot OH \qquad (39)$$

$$H\overset{\overset{O}{\|}}{C} \cdot + O_2 = HO_2 \cdot + CO \qquad (43)$$

The important features of these radical transfer reactions are the conversion of NO to NO_2 and the generation of OH radicals. As will be discussed later, the OH radicals are highly reactive with propylene and constitute the single most important chain carrier in the oxidation of propylene and NO. These transfer steps generally involve radicals of RO and RO_2 types where R is either hydrogen or an alkyl group for which the kinetics and mechanism are not well established. Particularly important in explaining the propylene chemistry are Reactions 6 and 39. The values of the rate constants for these two reactions are not known, but both have been assigned here to be equal to 2.0×10^{-13} cm^3 molecule^{-1} sec^{-1}. This assignment is based on numerical analysis of the data to give proper weight to these two reactions compared with other reactions involving these RO_2 radicals and is judged to be a lower limit for both rate constants. Lower values for these rate constants result in a marked deviation of the calculated profiles from the experimental data of Altshuller (*12, 13, 14*) and of other studies (*15, 16, 17, 18, 19, 20, 21, 22, 23, 24*).

Several termination reactions involving RO and RO_2 compete with the radical transfer steps. These reactions are grouped into $RO_2 + RO_2$, RO + NO, and RO + NO_2 types and are listed below.

$$CH_3CH_2O \cdot + NO_2 = CH_3CH_2ONO_2 \qquad (11)$$

$$2\, HO_2 \cdot = H_2O_2 + O_2 \qquad (40)$$

$$\cdot OH + NO = HONO \qquad (41)$$

$$\cdot OH + NO_2 = HNO_3 \qquad (42)$$

Published values of the rate constants for Reactions 40 and 41 (*29, 45*) were used in the numerical analysis. For the other termination reactions involving RO and RO_2 in this model, rate constants are not known. Estimated values were used that were derived to be consistent with k_{40} and k_{41}. The concentration of these radicals although always small in-

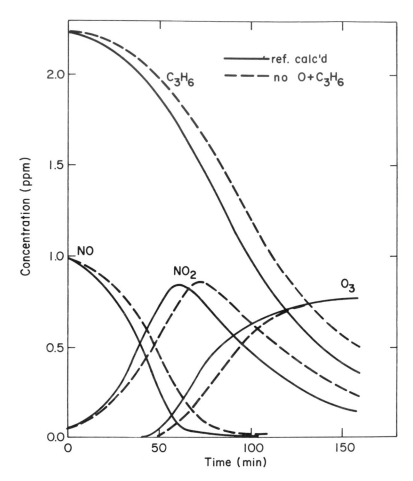

Figure 4. Effect of chain initiation by O-atom. A calculation with
$k_{4a} = 0$ *(dashed lines) is compared with the reference calculation,* k_{4a}
$= 1.2 \times 10^{-12}$ *cm³ molecule⁻¹ sec⁻¹ (solid lines).*

creases with time, and therefore reactions of the type $RO_2 + RO_2$ and
$RO + NO_2$ become progressively more important with time.

The significance of O-atom reactions is assessed in Figure 4 where
the effect of leaving Reaction 4a out of the overall kinetic mechanism is
shown. The induction period is extended without altering the overall
reaction profile. This agrees with the qualitative aspects of the O-atom
role discussed previously, using the simplified model. Reaction 4a is
significant only in the induction period when reaction chains are being
initiated. The system then sustains itself without further need of chain
initiation by O-atoms.

Oxidation of Propylene by Ozone. Hydrocarbon ozonolysis is important in smog chemistry, but the details of this ozone chemistry are still poorly understood despite numerous investigations (*46, 47, 48, 49, 50, 51, 52, 53, 54, 55, 56*). However, the ozonolysis of propylene is fairly well established as compared with the ozonolysis of internally bonded olefins. For example, the rate constant for propylene is uncertain by about 10% while for *trans*-2-butene the reported rate constants vary by an order of magnitude (*48, 49, 52, 55*). This provided another reason to choose propylene as the prototype hydrocarbon in this analysis.

The primary step of the propylene ozonolysis was taken to be the generally accepted Criegee mechanism (*46, 47*), which leads to the formation of zwitterions and aldehydes:

$$O_3 + CH_3CH=CH_2 \quad\begin{array}{l} \longrightarrow HC^+HOO^- + CH_3CHO \qquad (13a) \\ \longrightarrow CH_3C^+HOO^- + HCHO \qquad (13b) \end{array}$$

The rate constants for Reactions 13a and 13b given in Table AI were derived from values for the overall rate constant (*48, 49, 50, 51, 52, 53, 54, 55*) and the relative yields of formaldehyde and acetaldehyde (*51*). These relative yields of formaldehyde and acetaldehyde are consistent with the data of Altshuller *et al.*

According to the Criegee mechanism, formyl and acetyl zwitterions, HC^+HOO^- and $CH_3C^+HOO^-$, are reactive intermediates in the ozonolysis of propylene. Among several thermochemically feasible reactions of the zwitterions, the following scheme provides a chain oxidation mechanism which is consistent with the observed data,

$$HC^+HOO^- \quad + O_2 \quad = \quad \cdot OH \quad + \overset{\overset{\displaystyle O}{\|}}{H}COO\cdot \qquad (14)$$

$$CH_3C^+HOO^- + O_2 \quad = \quad \cdot OH \quad + CH_3\overset{\overset{\displaystyle O}{\|}}{C}OO\cdot \qquad (15)$$

$$\overset{\overset{\displaystyle O}{\|}}{H}COO\cdot \quad + NO \quad = \quad NO_2 \quad + \overset{\overset{\displaystyle O}{\|}}{H}CO\cdot \qquad (16)$$

$$\overset{\overset{\displaystyle O}{\|}}{H}COO\cdot \quad + NO_2 = \quad NO_3 \quad + \overset{\overset{\displaystyle O}{\|}}{H}CO\cdot \qquad (17)$$

$$\overset{\overset{\displaystyle O}{\|}}{H}CO\cdot \qquad\qquad = \quad H\cdot \quad + CO_2 \qquad (18)$$

$$H \cdot + O_2 \quad + M \quad = HO_2 \cdot + M \tag{19}$$

$$\underset{\|}{\overset{O}{CH_3C}}OO \cdot \quad + NO \quad = NO_2 \quad + \underset{\|}{\overset{O}{CH_3C}}O \cdot \tag{20}$$

$$2\underset{\|}{\overset{O}{CH_3C}}OO \cdot \quad\quad = 2\underset{\|}{\overset{O}{CH_3C}}O \cdot + O_2 \tag{23}$$

$$\underset{\|}{\overset{O}{CH_3C}}O \cdot \quad\quad = CO_2 \quad\quad + CH_3 \cdot \tag{24}$$

$$CH_3 \cdot \quad + O_2 \quad = CH_3OO \cdot \tag{25}$$

$$CH_3OO \cdot \quad + NO \quad = NO_2 \quad\quad + CH_3O \cdot \tag{26}$$

$$2CH_3OO \cdot \quad\quad = 2CH_3O \cdot + O_2 \tag{27}$$

$$CH_3O \cdot \quad + O_2 \quad = \underset{\|}{\overset{O}{HC}}H \quad\quad + HO_2 \cdot \tag{29}$$

$$HO_2 \cdot \quad + NO \quad = \cdot OH \quad\quad + NO_2 \tag{39}$$

Radical termination reactions which are unique to the ozonolysis are given by

$$\underset{\|}{\overset{O}{CH_3C}}OO \cdot \quad + NO_2 = PAN \ (\underset{\|}{\overset{O}{CH_3C}}OONO_2) \tag{21}$$

$$\underset{\|}{\overset{O}{CH_3C}}OO \cdot \quad + HO_2 \cdot = \underset{\|}{\overset{O}{CH_3C}}OOH + O_2 \tag{22}$$

$$CH_3OO \cdot \quad + HO_2 \cdot = CH_3OOH + O_2 \tag{28}$$

$$CH_3O \cdot \quad + NO \quad = CH_3ONO \tag{30}$$

$$CH_3O \cdot \quad + NO_2 = CH_3ONO_2 \tag{31}$$

Peroxyacetylnitrate (PAN) is a well-known product of the propylene ozonolysis in the presence of NO_2 and O_2 (57); this is strong evidence for the formation of peroxyacetyl (or acetyl) radicals in the subsequent reaction of acetyl zwitterion as given by Reaction 15. However, the peroxyacetyl radicals lead to the formation of PAN only after the NO–

NO₂ conversion is virtually completed. Thus, the competitive reactions of peroxyacetyl radicals with NO (Reaction 20) and with NO₂ (Reaction 21) were postulated, and their relative rate constants were chosen to match the experimental data.

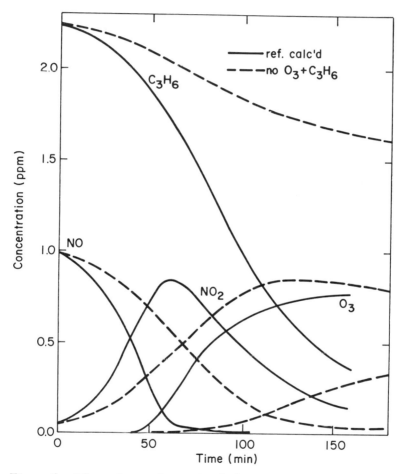

Figure 5.   Effect of ozonolysis.   A calculation with $k_{13} = 0$ *(dashed lines) is compared with the reference calculation,* $k_{13a} = 5.4 \times 10^{-18}$, $k_{13b} = 6.5 \times 10^{-18}$ *cm³ molecule⁻¹ sec⁻¹ (solid lines).*

The importance of the ozone reactions in all aspects of the mechanism is shown in Figure 5 where the behavior of the system without Reactions 13a and 13b is plotted. Radical production by ozonolysis represents an important part of the chain initiation throughout the entire course of reaction. The time of the induction period and the time of conversion of NO to NO₂ are lengthened by deleting these reactions.

The rate of C_3H_6 consumption after the NO_2 maximum is markedly reduced because the major loss processes during this stage are direct ozonolysis and ozonolysis-initiated chain reactions. The propylene decay and ozone formation therefore depart qualitatively and quantitatively from the experimental data.

Chain Oxidation of Propylene by OH Radicals. The potential role of OH radicals in explaining the excess rate of olefin consumption was discussed by Leighton in his monograph in 1961 (25), but no definite assessment of the importance was possible at that time. A quantitative analysis of the crucial role of an OH chain in explaining the excess rate during the NO–NO_2 conversion was made by the authors in 1969 (11). This was an early report of our analysis of the data discussed here. Heicklen presented an independent discussion of this question (6).

In the previous discussion of O-atom and O_3 reactions, the hydroxyl radical is a prominent product. As will be seen later, OH radicals are also produced in the photolyses of aldehydes and nitrous acid. The resulting OH chain mechanism for propylene is summarized below:

$$\cdot OH \quad + CH_3CH{=}CH_2 = CH_3\overset{\cdot}{C}H{-}CH_2OH \tag{32}$$

$$CH_3\overset{\cdot}{C}H{-}CH_2OH + O_2 = CH_3\overset{\overset{OO\cdot}{|}}{C}H{-}CH_2OH \tag{33}$$

$$CH_3\overset{\overset{OO\cdot}{|}}{C}H{-}CH_2OH + NO = NO_2 + CH_3\overset{\overset{O\cdot}{|}}{C}HCH_2OH \tag{34}$$

$$2CH_3\overset{\overset{OO\cdot}{|}}{C}HCH_2OH = 2CH_3\overset{\overset{O\cdot}{|}}{C}HCH_2OH + O_2 \tag{35}$$

$$CH_3\overset{\overset{OO\cdot}{|}}{C}HCH_2OH \quad + HO_2\cdot = CH_3\overset{\overset{OOH}{|}}{C}HCH_2OH + O_2 \tag{36}$$

$$CH_3\overset{\overset{O\cdot}{|}}{C}H{-}CH_2OH = CH_3\overset{\overset{O}{||}}{C}H + \cdot CH_2OH \tag{37}$$

$$\cdot CH_2OH \quad + O_2 = H\overset{\overset{O}{||}}{C}H + HO_2\cdot \tag{38}$$

$$HO_2\cdot \quad + NO = NO_2 + \cdot OH \tag{39}$$

Several other termination reactions involving HO_2 and OH have been discussed above.

The sensitivity of the reaction kinetics to the OH chain is illustrated in Figures 6 and 7 where the rate constant for Reaction 32 is taken to be half of the value given in Table AI and twice that value. The NO–NO_2 conversion and the concomitant consumption of propylene and formation of ozone are markedly affected. The effect of leaving the OH chain entirely out of the mechanism was discussed above with the simplified mechanism.

Oxidation of Aldehydes. Formaldehyde and acetaldehyde are major oxidation products of propylene that participate in smog reactions in

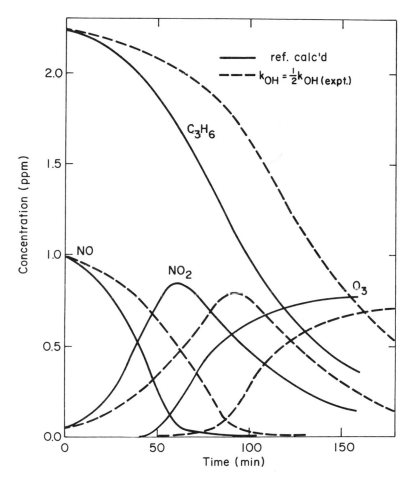

Figure 6.   Effect of reaction of OH with C_3H_6.   A calculation with $k_{32} = 8.5 \times 10^{-12}$ (dashed lines) is compared with the reference calculation, $k_{32} = 1.7 \times 10^{-11}$ cm^3 molecule^{-1} sec^{-1} (solid lines).

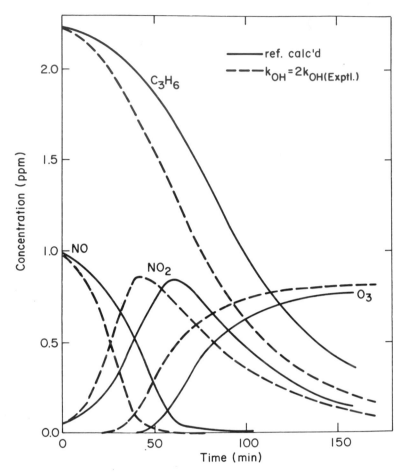

Figure 7. Effect of oxidation of C_3H_6 by OH. A calculation with k_{32} $= 3.4 \times 10^{-11}$ (dashed lines) is compared with the reference calculation, $k_{32} = 1.7 \times 10^{-11}$ cm³ molecule⁻¹ sec⁻¹ (solid lines).

numerous ways. For example, these aldehydes undergo photodissociation at wavelengths longer than 3000 A to provide a photochemical source of chain carriers:

$$\overset{O}{\overset{\|}{H C H}} + h\nu = H \cdot + \overset{O}{\overset{\|}{H C}} \cdot \qquad (56a)$$

$$\overset{O}{\overset{\|}{CH_3CH}} + h\nu = CH_3 \cdot + \overset{O}{\overset{\|}{HC}} \cdot \qquad (57)$$

The photodissociation rate of these steps was not derived for this experiment because spectral distribution data of the light source used was not reported. The values given in Table AI for 56a and 57 are 1% of the photodissociation rate of NO_2 in this system according to Leighton's estimates for sunlight (25). In Figure 8 the effect of 56a and 57 on the system is shown. The chain initiation stage of the reaction is extended, similarly to Figure 4 when the reaction $O + C_3H_6$ was omitted. If aldehydes were present as an initial constituent of the system, the time for the NO–NO_2 conversion would be decreased. However, this prediction should await experimental verification.

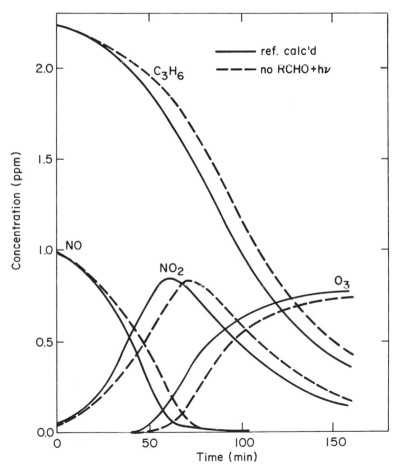

Figure 8.   Effect of aldehyde photolysis. A calculation with $k_{56a} = k_{57} = 0$ (dashed lines) is compared with the reference calculation, $k_{56a} = k_{57} = 6.7 \times 10^{-6}$ sec^{-1} (solid lines).

Also these aldehydes react with atomic oxygen and with OH radicals with rate constants that are roughly comparable with those of the reactant propylene.

$$O \; + \; H\overset{\overset{\displaystyle O}{\|}}{C}H \;\; = \; \cdot OH \; + \; H\overset{\overset{\displaystyle O}{\|}}{C} \cdot \qquad (43)$$

$$\cdot OH \; + \; H\overset{\overset{\displaystyle O}{\|}}{C}H \;\; = \; H_2O \; + \; H\overset{\overset{\displaystyle O}{\|}}{C} \cdot \qquad (44)$$

$$O \; + \; CH_3\overset{\overset{\displaystyle O}{\|}}{C}H \; = \; \cdot OH \; + \; CH_3\overset{\overset{\displaystyle O}{\|}}{C} \cdot \qquad (45)$$

$$\cdot OH \; + \; CH_3\overset{\overset{\displaystyle O}{\|}}{C}H \; = \; H_2O \; + \; CH_3\overset{\overset{\displaystyle O}{\|}}{C} \cdot \qquad (46)$$

The products of these reactions, OH, HCO, and CH_3CO, have been previously shown to be chain carriers, and accordingly the aldehydes will sustain chain reactions. One difference between these aldehydes and propylene is their relative rate of reaction with ozone, their rate being negligible compared with that of propylene.

Inorganic Reactions. Photooxidation of propylene in the presence of oxides of nitrogen involves numerous inorganic reactions. The role of the NO_2 photolysis in initiating O- and O_3-reactions has already been discussed. Another inorganic compound of photochemical interest is nitrous acid, HONO, since it provides another source for OH radicals. A reaction scheme for the formation and subsequent photodissociation of HONO is given by:

$$NO \quad + NO_2 + H_2O = HONO + HONO \qquad (53)$$

$$HONO + HONO \quad = NO \quad + NO_2 + H_2O \qquad (54)$$

$$HONO + h\nu \quad = NO \quad + \cdot OH \qquad (58)$$

Reactions 53, 54, and 58 are not well established, and the rate constants given in Table AI are somewhat speculative. For example, Reaction 53 is presumably a termolecular gas phase reaction of NO, NO_2, and H_2O (62), but the possibility of a heterogeneous mechanism cannot be entirely eliminated. The photo-dissociation rate for Reaction 58 (63, 64) is probably an upper limit, based on available experimental data. In any case, with the rate constants given in Table AI, exclusion of these three reactions from the mechanism would alter little the reaction profiles. Even if the equilibrium [HONO] were present, the reaction profiles would not be significantly changed.

Another series of inorganic reactions initiated by the reaction of O_3 with NO_2 are:

$$O_3 \quad + NO_2 = NO_3 \quad + O_2 \tag{48}$$

$$NO_3 + NO = NO_2 \quad + NO_2 \tag{49}$$

$$NO_3 + NO_2 = N_2O_5 \tag{50}$$

$$N_2O_5 \qquad = NO_2 \quad + NO_3 \tag{51}$$

$$N_2O_5 + H_2O = HNO_3 + HNO_3 \tag{52}$$

These reactions provide a mechanism for the consumption of NO_2 and concomitant production of HNO_3 in the later stages of the smog reaction. In this system an equilibrium concentration of N_2O_5 is maintained by Reactions 50 and 51. The formation rate of HNO_3 *via* Reaction 52 was estimated from the material balance for nitrogen-containing compounds in these experiments. To do this, the loss of NO_x was attributed to the PAN formation by Reaction 21 and HNO_3 by Reactions 42 and 52. The present estimate for the rate constant of Reaction 52 is several orders of magnitude smaller than previous estimates. The value of 1.7×10^{-18} derived by Jaffe and Ford (65) that is often cited does not agree with the data of this experiment. Presumably, the discrepancy between the value assigned here and that of Jaffe and Ford may result from the influence of wall reactions.

The overall effect of humidity on the smog reactions of propylene has been estimated, based on Reactions 48–54 and 58. Figure 9 shows a comparison of reaction profiles computed for zero and 50% relative humidity. According to the present kinetic model, the humidity effect seems to be rather slight. However, this prediction should be further verified experimentally. Existing smog chamber results on the humidity effect are not definitive and are frequently conflicting (66, 67, 68).

The chemical role of CO in smog reactions is examined by the following OH chain:

$$\cdot OH \quad + CO \qquad = CO_2 \quad + H \cdot \tag{55}$$

$$H \cdot \quad + O_2 + M = HO_2 \cdot + M \tag{19}$$

$$HO_2 \cdot + NO \qquad = NO_2 \quad + \cdot OH \tag{39}$$

The rate constant for the reaction of OH with CO is about two orders of magnitude smaller than that for the reaction of OH with propylene. Hence, the smog reactivity of CO based on the NO–NO$_2$ conversion rate should be roughly two orders of magnitude smaller than that of propylene. The kinetic role of CO was examined more quantitatively by calculating the effect of the addition of high concentrations of CO to the C$_3$H$_6$–NO$_x$ system. A typical result is shown in Figure 10 where 100 ppm CO was added to the system. The addition of CO shows two characteristic effects. The NO–NO$_2$ conversion in the initial stages of the reaction is accelerated, but in the later stages O$_3$ formation

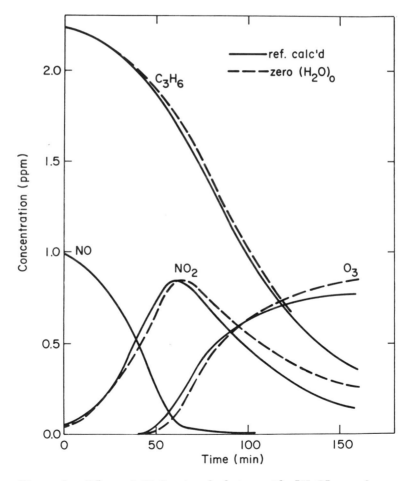

Figure 9. Effect of H$_2$O. A calculation with $[H_2O]_0 = 0$ ppm (dashed lines) is compared with the reference calculation, $[H_2O]_0 = 2.2 \times 10^4$ ppm $=50\%$ relative humidity (solid lines).

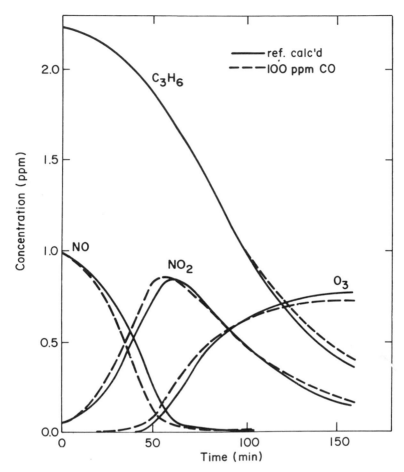

Figure 10. Effect of addition of CO. A calculation with $[CO]_0 = 100$ ppm (dashed lines) is compared with the reference calculation, $[CO]_0 = 0$ ppm (solid lines).

and consumption of propylene and NO_2 is reduced. This occurs because the chemical fate of the HO_2 radicals changes during the course of the smog reaction. In the early stages of the reaction, HO_2 is effectively removed by Reaction 39 to propagate the OH chain while in the later stages HO_2 is largely removed by the termination Reaction 40 and does not contribute to the regeneration of chain carriers. This predicted effect of CO is consistent with some experimental data (69, 70). However, in this analysis CO was assumed to react only with OH and not with other RO and RO_2 radicals. Although this assumption is plausible, further experimentational verification is desirable.

Discussion

Photochemical reactions in smog chambers have been studied to understand the generation of photochemical smog in the atmosphere, particularly in the Los Angeles Basin. The smog chamber studies have accordingly been designed to simulate atmospheric conditions as closely as possible. The results of these studies have been used as a basis of abatement strategy for the alleviation of smog although there have been a number of experimental limitations to this approach. The present analysis departs from this method. Instead, the experimental data have been used to derive a detailed mechanism for the chemical behavior of the propylene–NO_x system to provide a framework to analyze the more complicated atmospheric problem. From this analysis a number of critical elementary reactions have been identified for which quantitative information about the kinetics and mechanism is lacking. One aspect of our research program has been directed toward fulfilling this need. Where this information is lacking, estimates have been made that are consistent with the overall experimental behavior but which are not intended to be mere parametric fitting of the data. Perhaps the weakest area of our knowledge is that of the ozone chemistry. This is somewhat unfortunate because the main focus of abatement strategy has been on ozone. Some of the key questions are discussed below in terms of the detailed mechanism and its relation to the atmospheric chemistry.

Effect of Light Intensity. The primary effect of light intensity in this system is on the photodissociation of NO_2, HONO, and aldehydes. Of these the photolysis of NO_2 is the most important. As was shown in Equations III and IV, the concentrations of the two major chain initiators, O and O_3, are directly proportional to the light intensity. The O-atom concentration is also proportional to the NO_2 concentration, but the O_3 concentration is, more complicatedly, also proportional to $[NO_2]/[NO]$. Figure 11 shows the marked effect of reducing the light intensity by a factor of 2 for the detailed mechanism developed here. However, in the early stages of the reaction the effect is nearly linear, and the reaction profiles would nearly superimpose if the time scale were reduced by a factor of two. Kinetic indices, such as the half-time and the maximum rate for NO–NO_2 conversion that are most commonly used to describe smog reactivity, show a linear dependence on light intensity agreeing with this calculation because they stress the early stage of the reaction. There is some departure from linearity if one considers the half-time for propylene consumption, which extends beyond the initial reaction stage. More significantly, the terminal ozone concentration is decreased by much less than a factor of two. This is expected because as shown by Equation IV, the O_3 concentration is also a function of $[NO_2]/[NO]$, and this ratio is a complex function of the light intensity.

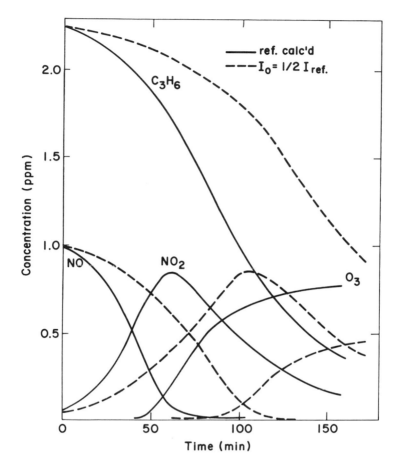

Figure 11. Effect of light intensity. A calculation with the light intensity reduced by a factor of 2 from the reference value (dashed lines) is compared with the reference calculation (solid lines).

Furthermore, the approximation used to derive Equation IV from Equation II, k_{O_3} [C_3H_6] $<< k_3$ [NO], is no longer valid for evaluating the terminal ozone concentration. Since the terminal ozone concentration is defined by the Air Quality Standard (*122*) as the index of smog severity, these considerations are particularly important in modeling atmospheric data.

Other factors relating to the light source that are significant in the application of this model to atmospheric conditions are the diurnal variation of intensity and spectral distribution. This should also add a complex nonlinear aspect to the analysis.

Finally, the absolute value of k_1 of 0.4 min^{-1}, used for the photodissociation of NO_2 in this analysis is uncertain to about 30%. This uncertainty has been previously discussed (22, 57) and arises from uncertainties in the measurement of k_d and in the factor relating k_1 to k_d. The major effect that this would have on this analysis would be to change the value estimated for k_{39}, the rate constant for the reaction of HO_2 with NO.

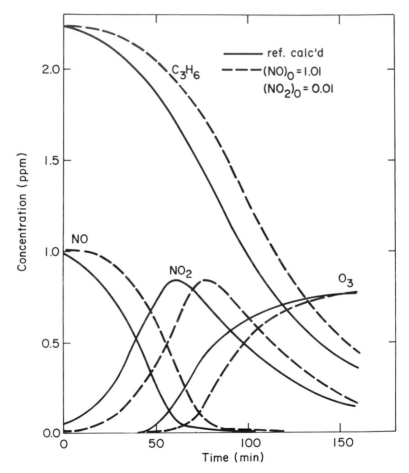

Figure 12. Effect of initial NO_2 concentration. A calculation with $[NO_2]_0 = 0.01$ ppm (dashed lines) is compared with the reference calculation, $[NO_2]_0 = 0.05$ ppm (solid lines).

Effect of Initial NO₂ Concentration. In laboratory studies conducted to simulate atmospheric photochemical smog, a serious effort is usually made to start the experiment with low initial concentrations of NO_2. In the data of Altshuller *et al.* studied here $[NO_2]_0$ was 0.05 ppm; in other studies $[NO_2]_0$ had lower values (15, 16, 17, 18, 19, 20), and in most of the studies reported $[NO_2]_0$ was not even specified. The effect of reducing $[NO_2]_0$ from 0.05 to 0.01 ppm is shown in Figure 12. The induction period is increased significantly because a reduction in $[NO_2]_0$ results in lower O-atom and O_3 concentrations (Equations III and IV) and consequently decreased chain initiation rates. A qualitatively similar effect was seen in Figure 4 when the O-atom reaction with propylene (Reaction 4) was omitted from the reaction scheme. As mentioned earlier, since most interpretations of smog chamber data generally emphasize the first stage of the reaction, consideration of this effect of $[NO_2]_0$ would resolve some discrepancies found by different investigators. For example, the oxidation rate of propylene found by Glasson and Tuesday (17) in a study similar to that analyzed here (13) differed significantly from that of Altshuller *et al.* This difference could be largely explained by the difference in $[NO_2]_0$ used in the two experiments (11).

Atmospheric conditions with respect to $[NO_2]_0$ are quite different from the laboratory experimental conditions. In polluted atmospheres the NO_2 to NO ratio is never as low as used in this experiment, and on days of severe smog the early morning concentrations of NO_2 are often equal to that of NO (71). The induction period, therefore, is less important under real atmospheric conditions than in laboratory experiments. Since many of the parameters describing laboratory results give great weight to the induction period, the application of these parameters to the interpretation of atmospheric data requires more detailed analysis than has previously been done.

Effect of NO Concentration. Since NO is the major initial constituent of NO_x, a variation in its concentration should affect a number of kinetic parameters and give rise to complex effects on the smog reactions. To illustrate the net effect of a variation in $[NO]_0$, a reaction profile has been computed by reducing $[NO]_0$ to 0.5 ppm from the reference value of 0.97 ppm with $[NO_2]_0$ and $[C_3H_6]_0$ unchanged. The results are shown in Figure 13. The most striking effect is a substantial reduction of the NO–NO_2 conversion time, as has also been observed experimentally (13, 15, 16, 17, 18, 19, 20). This reduction derives from two factors. The first is that, despite the decreased concentration of NO, the rate of conversion to NO_2 is roughly the same as in the reference calculation. This unusual behavior is primarily the result of higher O_3 concentrations, which are governed by the ratio, $[NO_2]/[NO]$ (Equation IV). Having the same rate and half as much NO to convert to NO_2, the conversion time is about cut in half. Also, propylene consumption is enhanced, another consequence of the earlier rise in O_3 concentration.

The more rapid NO conversion and propylene consumption that result from a decrease in the initial NO concentration have been cited as an index of increased smog severity with the corollary conclusion that a decrease in NO concentration in urban atmospheres may result in

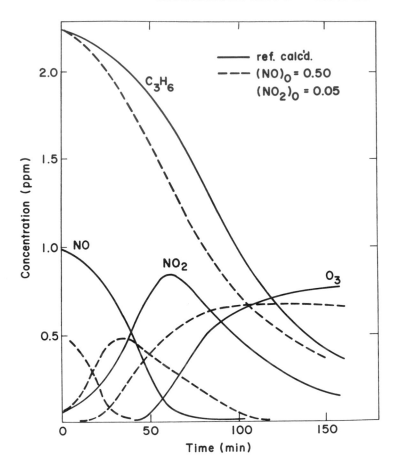

Figure 13.   Effect of initial NO concentration. A calculation with [NO]$_0$ = 0.50 *ppm (dashed lines) is compared with the reference calculation,* [NO]$_0$ = 0.97 *ppm (solid lines).*

greater smog intensity (*15*). However, this behavior is largely influenced by changes that occur in the early stages of the reaction. Since the early stages of the reaction are less important under atmospheric conditions, this conclusion may not be entirely valid. In Figure 13 an example of the questionable validity of this conclusion is shown. Despite the fact that O_3 builds up more rapidly with the lower initial concentration of NO, its peak value is lower than that in the reference experiment. Furthermore, although not shown in Figure 13, the concentrations of nitrogen containing smog products, such as PAN and nitric acid, are also reduced.

Effect of C_3H_6 Concentration. Variation of the propylene concentration gives rise to complex changes in the reaction parameters that are similar to those that resulted from a variation in the NO concentration.

To illustrate this, the reaction profile for a factor of two reduction in $[C_3H_6]$ without change in $[NO]_0$ and $[NO_2]_0$ is plotted in Figure 14. The time for conversion of NO to NO_2 is nearly doubled as is the time to consume half of the propylene. These parameters are characteristic of the early stages of the reaction where propylene participates in chain initiation by Reactions 4a and 13 and in chain propagation by Reaction 32. This predicted behavior agrees with smog chamber observations that pertain chiefly to the earlier stages of the reaction. (*13, 15, 16, 17, 18, 19, 20*). The later stage of the reaction is not shown because ∼ 300 minutes are required to reach the terminal ozone concentration. Here the effect of reducing the initial propylene concentration by a factor of two is not as pronounced. The value for this calculation is 0.64 ppm

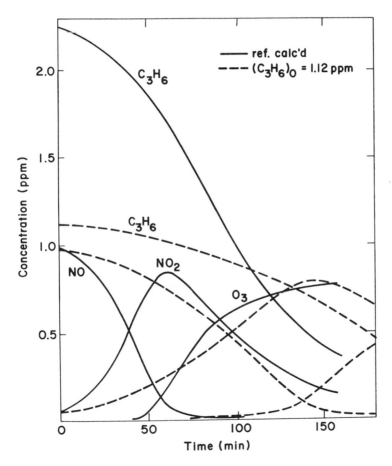

Figure 14. Effect of initial C_3H_6 concentration. A calculation with $[C_3H_6]_0 = 1.12$ ppm (dashed lines) is compared with the reference calculation, $[C_3H_6]_0 = 2.23$ ppm (solid lines).

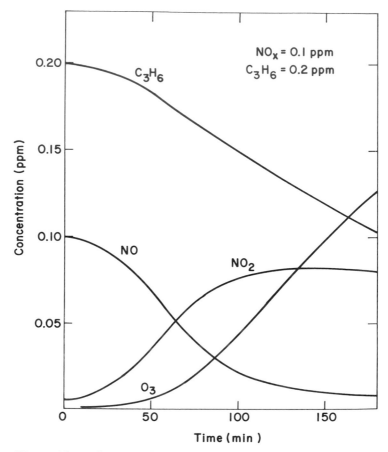

Figure 15. Effect of dilution. In this calculation, the initial con-
centrations of propylene (0.2 ppm), nitric oxide (0.1 ppm), and
nitrogen dioxide (0.005 ppm) are reduced by a factor of 10 from
the reference concentrations.

compared with 0.76 for the reference calculation. The total amount of
nitrogen containing smog products are the same in the two systems.

This qualitative behavior is analogous to observations in the Los
Angeles Basin. There hydrocarbon levels have been reduced without
equivalent reduction in NO_x, and presumably ozone formation in the
vicinity of downtown Los Angeles has decreased while a similar decrease
has not occurred in areas downwind where the reacting air mass has
experienced a longer reaction time (72).

Effect of Dilution. A discussion of the variation of NO and C_3H_6
concentrations separately is not complete without considering the effect
of their relative concentration or the $[C_3H_6]/[NO]$ ratio. In a complex
system of this type where there are multiple competing reactions, the

effect of dilution on the reacting species may be extremely nonlinear. Several calculations have been made on the effect of dilution on the reference system, keeping the ratio of $[C_3H_6]$ to $[NO_x]$ nearly constant. The extreme case for which calculations were made is shown in Figure 15 where the concentrations have been reduced by a factor of ten. The most significant effect of dilution is that the NO_2 consumption to form nitrogen containing smog products is substantially reduced while the time to reach the NO_2 maximum concentration is relatively unaffected. This latter behavior is a consequence of the fact that the initiation rates by Reactions 4a and 13 are proportional to the concentration of propylene while the corresponding rates for termination by Reactions 41 and 42 are proportional to the NO_x concentration. The chain length in this early stage of the reaction is then proportional to the ratio, $[C_3H_6]/[NO_x]$. Since the time to reach the maximum concentration of NO_2 depends largely on the OH chain, it is not greatly affected by dilution of C_3H_6 and of NO_x if the ratio of their concentrations is kept constant. One reason for the decrease in NO_2 consumption is that removal of NO_2 through ozonolysis and PAN formation is drastically reduced because

Table I. Relative Rate Constants and Reactivity

Compound	Relative Rate Constants			Reactivity [d]
	O [a]	O_3 [b]	OH [c]	
Olefins				
Ethylene	0.2	0.3	0.1	0.3
Isobutene	4.4	1.7~2	2.5	1
Trans-2-butene	4.9	2.8~36	4.2	4
2-Methyl-2-butene	14	2.4	7.1	6
Tetramethylethylene	18	3.1~62	8.6	17
Aromatics				
Benzene	.007	—	≤0.05	0.1
Xylene	—	<0.2	1.1	1
Aldehydes				
Formaldehyde	0.05	—	0.9	0.7
Acetaldehyde	.15	—	0.9	0.7
Propionaldehyde	~.2	—	1.8	2
Alkanes				
n-Butane	.008	—	0.24	0.2

[a] From References *33, 35, 79,* and *80,* relative to k_4.
[b] From Reference *29,* relative to k_{13}.
[c] From Reference *59,* relative to k_{32}.
[d] From References *20* and *121,* relative to propylene reactivity, based on the rate of conversion of NO to NO_2.

of the decreased concentrations of all of the reacting species. Another reason is that NO_2 must compete with molecular oxygen for the radical precursors to form nitrogen containing products—*e.g.*, in Reactions 11 and 9 and in 31 and 29.

The time required to consume half of the propylene for this tenfold dilution is increased by only a factor of two. This is because the halftime is largely expended in the early stage of the reaction, which is relatively unaffected. This shows further that the halftime for hydrocarbon consumption need not correlate with smog intensity.

The ozone formation rate is substantially reduced, but its terminal value at 300 minutes is only reduced by 50% from that derived for a tenfold greater concentration. This predicted insensitivity of the terminal ozone concentration to reactant concentrations suggests that ozone may be a poor index of air quality at low concentrations.

Effect of Other Hydrocarbons. The complex role of C_3H_6 in this system has been analyzed in terms of its reactions with O, O_3, and OH, and a similar analysis is applied to determining the role of other hydrocarbons in the smog chemistry. Relative rate constants for the reactions of a number of hydrocarbons with O, O_3, and OH hydrocarbons are presented in Table I, together with the relative reactivities of these hydrocarbons based on NO–NO_2 conversion rates observed in smog chambers. The relative OH rate constants correlate remarkably well with the reactivities for all the types of hydrocarbons listed in the table. By contrast, the O-atom and O_3 rates correlate with the reactivities only for the olefinic compounds. For the aromatics, aldehydes, and alkanes in the table, the relative O-atom and O_3 rate constants are negligibly small compared with the relative reactivities. The relative OH rate

Table AI.   Complete

Reaction

Photolysis of Nitrogen Dioxide
(1) $NO_2 + h\nu$ $\rightarrow NO + O$
(2) $O + O_2 + M$ $\rightarrow O_3 + M(=air)$
(3) $NO + O_3$ $\rightarrow NO_2 + O_2$

Oxidation of Propylene by O-Atom

(4a) $O + CH_3CH{=}CH_2$ $\rightarrow CH_3CH_2\cdot + H\overset{\overset{\displaystyle O}{\|}}{C}\cdot$

(4b) $\rightarrow CH_3CH_2\overset{\overset{\displaystyle O}{\|}}{C}H$

constants have only recently been measured (*59*) and explain one puzzling aspect of reactivity scales—*i.e.*, the poor correlation with O-atom and O_3 rate constants for many hydrocarbon classes.

As has been discussed in some detail, OH plays its principal role in the conversion of NO to NO_2 during the early stages of the reaction. With respect to the NO to NO_2 conversion, all of the hydrocarbon classes of Table I have more or less comparable reactivity. Therefore, the procedure used to evaluate the effective hydrocarbon concentration of atmospheric mixtures—*i.e.*, measuring the ppm concentration of nonmethane hydrocarbons, should give a reliable average hydrocarbon concentration for predicting the NO–NO_2 conversion rate. However, the later stage of the reaction which governs the terminal concentrations of ozone and nitrogen-containing smog products is dominated by ozonolysis reactions. Since the various hydrocarbon classes in Table I have markedly different rate constants for reaction with ozone, this simple averaging procedure should be inapplicable for predicting the levels of terminal products. For example, hydrocarbon abatement regulations that are already in force and that are planned for the future may result in an increased relative concentration of alkanes in urban areas where they are present as background hydrocarbons (*73, 74, 123*). While the general decrease in hydrocarbon concentration will result in decreased ozone formation, the terminal ozone concentration may be much less affected because the removal of ozone by reaction with hydrocarbons will be largely curtailed while its net formation by increase in the $[NO_2]/[NO]$ will be sustained by hydrocarbons such as butane that are unreactive with ozone. This form of synergism needs careful consideration to understand whether or not the Air Quality Standard for ozone is attainable.

Reaction List

Rate Constant [a]	*References*
6.6 (−3)	*15,21,22,25,121*
6.0 (−34)	*28,76,77*
2.0 (−14)	*21,22,28,29,121*
1.2 (−12)	*25,29,32,33,34,35,36,37,38,39,40,41,42,43,44*
8.0 (−13)	*25,29,32,33,34,35,36,37,38,39,40,41,42,43,44*

Reaction

(4c) $\rightarrow CH_3\overset{\displaystyle O}{\overset{\displaystyle \diagup \diagdown}{CH-CH_2}}$

(5) $CH_3CH_2 \cdot + O_2$ $\rightarrow CH_3CH_2OO \cdot$

(6) $CH_3CH_2OO \cdot + NO$ $\rightarrow NO_2 + CH_3CH_2O \cdot$

(7) $2CH_3CH_2OO \cdot$ $\rightarrow 2CH_3CH_2O \cdot + O_2$

(8) $CH_3CH_2OO \cdot + HO_2 \cdot$ $\rightarrow CH_3CH_2OOH + O_2$

(9) $CH_3CH_2O \cdot + O_2$ $\rightarrow CH_3\overset{\displaystyle O}{\overset{\|}{C}}H + HO_2 \cdot$

(10) $CH_3CH_2O \cdot + NO$ $\rightarrow CH_3CH_2ONO$

(11) $CH_3CH_2O \cdot + NO_2$ $\rightarrow CH_3CH_2ONO_2$

(12) $H\overset{\displaystyle O}{\overset{\|}{C}} \cdot + O_2$ $\rightarrow HO_2 \cdot + CO$

Oxidation of Propylene by Ozone

(13a) $O_3 + CH_3CH=CH_2$ $\rightarrow H\overset{+}{C}H\overset{-}{OO} + CH_3\overset{\displaystyle O}{\overset{\|}{C}}H$

(13b) $\rightarrow CH_3\overset{+}{C}H\overset{-}{OO} + H\overset{\displaystyle O}{\overset{\|}{C}}H$

(14) $H\overset{+}{C}H\overset{-}{OO} + O_2$ $\rightarrow \cdot OH + H\overset{\displaystyle O}{\overset{\|}{C}}OO \cdot$

(15) $CH_3\overset{+}{C}H\overset{-}{OO} + O_2$ $\rightarrow \cdot OH + CH_3\overset{\displaystyle O}{\overset{\|}{C}}OO \cdot$

(16) $H\overset{\displaystyle O}{\overset{\|}{C}}OO \cdot + NO$ $\rightarrow NO_2 + H\overset{\displaystyle O}{\overset{\|}{C}}O \cdot$

(17) $H\overset{\displaystyle O}{\overset{\|}{C}}OO \cdot + NO_2$ $\rightarrow NO_3 + H\overset{\displaystyle O}{\overset{\|}{C}}O \cdot$

(18) $H\overset{\displaystyle O}{\overset{\|}{C}}O \cdot$ $\rightarrow H \cdot + CO_2$

(19) $H \cdot + O_2 + M$ $\rightarrow HO_2 \cdot + M$

(20) $CH_3\overset{\displaystyle O}{\overset{\|}{C}}OO \cdot + NO$ $\rightarrow NO_2 + CH_3\overset{\displaystyle O}{\overset{\|}{C}}O \cdot$

(21) $CH_3\overset{\displaystyle O}{\overset{\|}{C}}OO \cdot + NO_2$ $\rightarrow CH_3\overset{\displaystyle O}{\overset{\|}{C}}OONO_2$
                                              (PAN)

Continued

Rate Constant[a]	*References*
1.0 (−12)	*25,29,32,33,34,35,36,37,38,39,40,41,42,43,44*
>2.5 (−15)	*84,85,86,87,88,89,90,91,92,93*
2.0 (−13)	*102,103*
3.0 (−12)	*84,85,86,90,91,115*
3.6 (−12)	*84,85,86,90,91,115*
3.0 (−18)	*84,85,86,87,90,91*
6.7 (−13)	*103,114*
2.0 (−12)	*103*
>2.0 (−15)	*84,85,86,87,95*
5.4 (−18)	*21,22,25,29,46,47*
6.5 (−18)	*21,22,25,29,46,47*
1.0 (−18)	*21,22,25,46,47*
1.0 (−18)	*21,22,25,46,47*
1.0 (−12)	*57,102,103*
1.5 (−14)	*57,102,105*
1.0 (+12)	*104,105*
1.2 (−32)	*28,29,82,83*
1.0 (−12)	*102,103*
1.5 (−14)	*57,102,103,121*

Table AI.

Reaction

(22) $CH_3\overset{\displaystyle O}{\overset{\|}{C}}OO\cdot + HO_2\cdot \quad \rightarrow CH_3\overset{\displaystyle O}{\overset{\|}{C}}OOH + O_2$

(23) $2CH_3\overset{\displaystyle O}{\overset{\|}{C}}OO\cdot \quad \rightarrow 2CH_3\overset{\displaystyle O}{\overset{\|}{C}}O\cdot + O_2$

(24) $CH_3\overset{\displaystyle O}{\overset{\|}{C}}O\cdot \quad \rightarrow CH_3\cdot + CO_2$

(25) $CH_3\cdot + O_2 \quad \rightarrow CH_3OO\cdot$

(26) $CH_3OO\cdot + NO \quad \rightarrow CH_3O\cdot + NO_2$

(27) $2CH_3OO\cdot \quad \rightarrow 2CH_3O\cdot + O_2$

(28) $CH_3OO\cdot + HO_2\cdot \quad \rightarrow CH_3OOH + O_2$

(29) $CH_3O\cdot + O_2 \quad \rightarrow HO_2\cdot + H\overset{\displaystyle O}{\overset{\|}{C}}H$

(30) $CH_3O\cdot + NO \quad \rightarrow CH_3ONO$

(31) $CH_3O\cdot + NO_2 \quad \rightarrow CH_3ONO_2$

OH–Propylene Chain

(32) $\cdot OH + CH_3CH{=}CH_2 \rightarrow CH_3\overset{\displaystyle \cdot}{C}HCH_2OH$

(33) $CH_3\overset{\displaystyle \cdot}{C}HCH_2OH + O_2 \rightarrow CH_3\underset{\displaystyle OO\cdot}{C}HCH_2OH$

(34) $CH_3\underset{\displaystyle OO\cdot}{C}HCH_2OH + NO \rightarrow NO_2 + CH_3\underset{\displaystyle O\cdot}{C}HCH_2OH$

(35) $2CH_3\underset{\displaystyle OO\cdot}{C}HCH_2OH \quad \rightarrow O_2 + 2CH_3\underset{\displaystyle O\cdot}{C}HCH_2OH$

(36) $CH_3\underset{\displaystyle OO\cdot}{C}HCH_2OH + HO_2\cdot\rightarrow CH_3\underset{\displaystyle OOH}{C}HCH_2OH + O_2$

(37) $CH_3\underset{\displaystyle O\cdot}{C}HCH_2OH \quad \rightarrow CH_3\overset{\displaystyle O}{\overset{\|}{C}}H + \cdot CH_2OH$

(38) $\cdot CH_2OH + O_2 \quad \rightarrow HO_2\cdot + H\overset{\displaystyle O}{\overset{\|}{C}}H$

(39) $HO_2\cdot + NO \quad \rightarrow NO_2 + \cdot OH$

(40) $2HO_2\cdot \quad \rightarrow HOOH + O_2$

(41) $\cdot OH + NO \quad \overset{(M)}{\rightarrow} HONO$

(42) $\cdot OH + NO_2 \quad \overset{(M)}{\rightarrow} HNO_3$

Continued

Rate Constant[a]	*References*
3.6 (-12)	*84,85,86,90,91,115*
3.0 (-12)	*84,85,86,90,91,115*
1.0 $(+12)$	*104,105*
$>2.5\ (-15)$	*84,85,86,87,88,89,90,91*
2.0 (-13)	*102,103*
3.0 (-12)	*84,85,86,90,91,115*
3.6 (-12)	*84,85,86,90,91,115*
3.0 (-18)	*84,85,86,87,90,91*
6.7 (-13)	*103,114*
2.0 (-12)	*103*
1.7 (-11)	*21,22,58,59,62,106,107,108,109*
2.0 (-15)	*84,85,86,87,88,89,90*
2.0 (-13)	*103,107*
3.0 (-12)	*84,85,86,90,91,115*
3.6 (-12)	*84,85,86,90,91,115*
1.0 $(+12)$	*84,85,86,90,91,115*
1.0 (-18)	*84,85,86,87,90,91*
2.0 (-13)	*29,98,99,100,101*
3.6 (-12)	*28,29,94,114*
1.4 (-12)	*98,99,100,110,111,112,113*
4.1 (-12)	*98,99,100,110,111,112,113*

Reaction

Formaldehyde Oxidation

$$(43) \quad O + H\overset{O}{\underset{||}{C}}H \quad \rightarrow \cdot OH + H\overset{O}{\underset{||}{C}}\cdot$$

$$(44) \quad \cdot OH + H\overset{O}{\underset{||}{C}}H \quad \rightarrow H_2O + H\overset{O}{\underset{||}{C}}\cdot$$

Acetaldehyde Oxidation

$$(45) \quad O + CH_3\overset{O}{\underset{||}{C}}H \quad \rightarrow \cdot OH + CH_3\overset{O}{\underset{||}{C}}\cdot$$

$$(46) \quad \cdot OH + CH_3\overset{O}{\underset{||}{C}}H \quad \rightarrow H_2O + CH_3\overset{O}{\underset{||}{C}}\cdot$$

$$(47) \quad CH_3\overset{O}{\underset{||}{C}}\cdot + O_2 \quad \rightarrow CH_3\overset{O}{\underset{||}{C}}OO\cdot$$

Inorganic Reactions

(48) $NO_2 + O_3$ $\rightarrow NO_3 + O_2$
(49) $NO + NO_3$ $\rightarrow 2NO_2$
(50) $NO_2 + NO_3$ $\rightarrow N_2O_5$
(51) N_2O_5 $\rightarrow NO_2 + NO_3$
(52) $N_2O_5 + H_2O$ $\rightarrow 2HNO_3$
(53) $NO + NO_2 + H_2O$ $\rightarrow 2HONO$
(54) $2HONO$ $\rightarrow NO + NO_2 + H_2O$
(55) $\cdot OH + CO$ $\rightarrow CO_2 + H\cdot$

Photolysis of Reaction Products

$$(56a) \quad H\overset{O}{\underset{||}{C}}H + h\nu \quad \rightarrow H\overset{O}{\underset{||}{C}}\cdot + H\cdot$$

$$(56b) \quad H\overset{O}{\underset{||}{C}}H + h\nu \quad \rightarrow H_2 + CO$$

$$(57) \quad CH_3\overset{O}{\underset{||}{C}}H + h\nu \quad \rightarrow CH_3\cdot + H\overset{O}{\underset{||}{C}}\cdot$$

(58) $HONO + h\nu$ $\rightarrow \cdot OH + NO$
(59) $CH_3ONO + h\nu$ $\rightarrow CH_3O\cdot + NO$
(60) $CH_3CH_2ONO + h\nu$ $\rightarrow CH_3CH_2O\cdot + NO$

^{a} Units of rate constants (300°K):
First order reactions—sec^{-1}; second order reactions—cm^3 molecule^{-1} sec^{-1}; and third order reactions—cm^6 molecule^{-2} sec^{-1}. (The number in parentheses refers to the

Continued

Rate Constant[a]	References
1.8 (−13)	43,79,80
1.6 (−11)	43,62,98,108
3.3 (−13)	79,80,81
1.6 (−11)	62,107,108
> 2.5 (−15)	84,85,86,87,90,91
7.2 (−17)	25,29
1.0 (−11)	25,29,112,116,121
3.0 (−12)	25,29,111,116,121
2.3 (−1)	25,29,116,117,121
1.0 (−21)	25,26,27,29,63,64
1.0 (−36)	25,26,63,64
1.9 (−17)	25,26,63,64
1.8 (−13)	28,29,82,83
6.7 (−6)	25,26,60,61,78
6.7 (−6)	25,26,60,61,78
6.7 (−6)	25,26,78
3.0 (−6)	25,65
3.0 (−6)	25,114
3.0 (−6)	25,114

power of ten by which the number outside the parentheses should be multiplied—*e.g.*, in Reaction 60, 3.0 (−6) = 3.0 × 10⁻⁶.)

Literature Cited

1. Eschenroeder, A. Q., Martinez, J. R., "Concepts and Applications of Photochemical Smog Reactions," General Research Corporation, Santa Barbara (June 1971).
2. Eschenroeder, A. Q., Martinez, J. R., "Mathematical Modeling of Photochemical Smog," General Research Corporation, Santa Barbara (December 1969).
3. Eschenroeder, A. Q., "Validation of Simplified Kinetics for Photochemical Smog Modeling," General Research Corporation, Santa Barbara (September 1969).
4. Hecht, T. A., Seinfeld, J. H., *Environ. Sci. Technol.* (1972) **6**, 47.
5. Friedlander, S. K., Seinfeld, J. H., *Environ. Sci. Technol.* (1969) **3**, 1175.
6. Heicklen, J., Westberg, K., Cohen, N., "Chemical Reactions in Urban Atmospheres," p. 55, C. S. Tuesday, Ed., American Elsevier, New York, 1971.
7. Westberg, K., Cohen, N., "The Chemical Kinetics of Photochemical Smog as Analyzed by Computer," The Aerospace Corp., El Segundo (December 1969).
8. Wayne, L. G., Danchick, R., Weisburd, M., Kokin, A., Stein, A., "Modeling Photochemical Smog on a Computer for Decision-Making," System Development Corporation, Santa Monica (September 1970).
9. Wayne, L. G., Ernest, T. E., "Photochemical Smog, Simulated by Computer," Paper No. 69-15, *Air Pollut. Control Assoc. Ann. Meet.*, New York, (June 1969).
10. Behar, J., "Simulation Model of Air Pollution Photochemistry," Vol. 4, Project Clean Air, University of California (September 1970).
11. Weinstock, B., Daby, E. E., Niki, H., "Chemical Reactions in Urban Atmospheres," p. 54, C. S. Tuesday, Ed., American Elsevier, New York, 1971.
12. Altshuller, A. P., personal communication, July 1969.
13. Altshuller, A. P., Kopczynski, S. L., Lonneman, W. A., Becker, T. L., Slater, R., *Environ. Sci. Technol.* (1967) **1**, 899.
14. Altshuller, A. P., Kopczynski, S. C., Wilson, D., Lonneman, W. A., Sutterfield, F. D., *J. Air Pollut. Control Assoc.* (1969) **19**, 787.
15. Glasson, W. A., Tuesday, C. S., *Environ. Sci. Technol.* (1971) **5**, 151.
16. Glasson, W. A., Tuesday, C. S., *J. Air Pollut. Control Assoc.* (1970) **20**, 239.
17. Glasson, W. A., Tuesday, C. S., *Environ. Sci. Technol.* (1970) **4**, 37.
18. Caplan, J. D., Society of Automotive Engineers, *Trans.* (1966) **74**, 197.
19. Tuesday, C. S., "Chemical Reactions in the Lower and Upper Atmosphere," p. 15, R. Cadle, Ed., Interscience, New York, 1961.
20. Glasson, W. A., Tuesday, C. S., *Environ. Sci. Technol.* (1970) **4**, 916.
21. Stedman, D. H., Morris, E. D., Jr., Daby, E. E., Niki, H., Weinstock, B., "The Role of OH Radicals in Photochemical Smog," *Amer. Chem. Soc. Nat. Meet., 160th*, Chicago (September 1970).
22. Stedman, D. H., Daby, E. E., Niki, H., Weinstock, B., "The Relative Effectiveness of Hydrocarbons in $NO-NO_2$ Conversion, and the Photostationary $NO-NO_2-O_3$ Equilibrium in Photochemical Smog," *Amer. Chem. Soc. Nat. Meet., 161st*, Los Angeles (March 1971).
23. Schuck, E. A., Doyle, G. S., Endow, N., Report No. 31, Air Pollution Foundation, San Marino (December 1960).
24. Hurn, R. W., Dimitriades, B., Fleming, R. D., Society of Automotive Engineers, Mid-year Meeting, Chicago (May 1965).
25. Leighton, P. A., "Photochemistry of Air Pollution," Academic, New York, 1961.

26. Altshuller, A. P., Bufalini, J. J., *Environ. Sci. Technol.* (1971) **5**, 39.
27. Altshuller, A. P., Bufalini, J. J., *Photochem. and Photobiol.* (1965) **4**, 97.
28. Schofield, K., *Planet. Space Sci.* (1967) **15**, 643.
29. Johnston, H. S., Pitts, J. N., Jr., Lewis, J., Zafonte, L., Mottershead, T., "Atmospheric Chemistry and Physics," Project Clean Air, Vol. 4, University of California (September 1970).
30. Gay, B. W., Bufalini, J. J., *Environ. Sci. Technol.* (1971) **5**, 422.
31. Niki, H., Warnick, A., Lord, R. R., *Trans.* (1972) **80**, 246.
32. Atkinson, R., Cvetanović, R. J., *J. Chem. Phys.* (1971) **55**, 659.
33. Cvetanović, R. S., *Advan. Photochem.* (1965) **1**, 115.
34. Cvetanović, R. J., *J. Air Pollut. Control Assoc.* (1964) **14**, 208.
35. Niki, H., Daby, E. E., Weinstock, B., Twelfth Symposium (International) on Combustion, p. 277 (1969).
36. Westenberg, A. A., de Haas, N., Twelfth Symposium (International) on Combustion, p. 289 (1969).
37. Brown, J. M., Thrush, B. A., *Trans. Faraday Soc.* (1967) **63**, 630.
38. Saunders, D., Heicklen, J., *J. Phys. Chem.* (1966) **70**, 1950.
39. Elias, L., Schiff, H. I., *Can. J. Chem.* (1960) **38**, 1657.
40. Elias, L., *J. Chem. Phys.* (1963) **38**, 989.
41. Azatyan, V. V., Nalbandyan, A. B., Meng-yuan, T., *Doklad, Izv. Nauk USSR* (1963) **149**, 1095.
42. Avramenko, L. I., Kolesnikova, R. J., *Advan. Photochem.* (1964) **2**, 25.
43. Herron, J. T., Penzhorn, R. D., *J. Phys. Chem.* (1969) **73**, 191.
44. Stuhl, F., Niki, H., *J. Chem. Phys.* (1971) **55**, 3954.
45. Foner, S. N., Hudson, R. L., *Advan. Chem. Ser.* (1962) **36**, 34.
46. Criegee, R., Blust, G., Zinke, H., *Chem. Ber.* (1964) **87**, 766.
47. Criegee, R., Kerchow, A., Zinke, H., *Chem. Ber.* (1955) **88**, 1878.
48. Wei, Y. K., Cvetanović, R. J., *Can. J. Chem.* (1963) **41**, 913.
49. Vrbaski, T., Cvetanović, R. J., *Can. J. Chem.* (1960) **38**, 1053, 1063.
50. Williamson, D. G., Cvetanović, R. J., *J. Amer. Chem. Soc.* (1968) **90**, 3668.
51. Hanst, P. L., Stephens, E. R., Scott, W. E., Doerr, R. C., "Atmospheric Ozone-Olefin Reactions," The Franklin Institute, Philadelphia, Pa., August 1958.
52. Bufalini, J. J., Altshuller, A. P., *Can. J. Chem.* (1965) **43**, 2243.
53. Cadle, R. D., Schadt, C., *J. Amer. Chem. Soc.* (1952) **74**, 6002.
54. DeMore, W. B., *Int. J. Chem. Kinetics* (1969) **1**, 209.
55. Schuck, E. A., Doyle, G. J., Report No. 29, Air Pollution Foundation, San Marino (October 1959).
56. Bailey, P. S., *Chem. Rev.* (1958) **58**, 925.
57. Stephens, E. R., *Advan. Environ. Sci.* (1970) **1**, 119.
58. Morris, E. D., Jr., Stedman, D. H., Niki, H., *J. Amer. Chem. Soc.* (1971) **93**, 3570.
59. Morris, E. D., Jr., Niki, H., *J. Chem. Phys.* (1971) **55**, 1991.
60. McQuigg, R. D., Calvert, J. G., *J. Amer. Chem. Soc.* (1969) **91**, 1590.
61. DeGraff, B. A., Calvert, J. B., *J. Amer. Chem. Soc.* (1967) **89**, 2247.
62. Wayne, L. G., Yost, D. M., *J. Chem. Phys.* (1951) **19**, 41.
63. Asquith, P. L., Tyler, B. J., *Chem. Commun.* (1970) **1970**, 744.
64. King, G. W., Moule, G., *Can. J. Chem.* (1962) **40**, 2057.
65. Jaffe, S., Ford, H. W., *J. Phys. Chem.* (1967) **71**, 1832.
66. Dimitriades, B., *J. Air Pollut. Control Assoc.* (1967) **17**, 460.
67. Bufalini, J. J., Altshuller, A. P., *Environ. Sci. Technol.* (1969) **3**, 469.
68. Wilson, W. E., Jr., Levy, A., "The Effect of Water Vapor on the Oxidation of 1-Butene and NO in the Photochemical Smog Reaction," *Amer. Chem. Soc. Nat. Meet., 157th,* Minneapolis (April 1969).
69. Westberg, K., Cohen, N., Wilson, K. W., *Science* (1971) **171**, 1013.

70. Wilson, W. E., Ward, G. F., "The Role of Carbon Monoxide in Photochemical Smog. I. Experimental Evidence for Its Reactivity," *Amer. Chem. Soc. Nat. Meet., 160th,* Chicago (September 1970).
71. "Final Report, 1969 Atmospheric Reaction Studies in the Los Angeles Basin," Vol. I-IV, Scott Research Laboratories, February 1970, prepared for CRC and NAPCA.
72. California Air Resources Board Bulletin (July–September 1971).
73. Stephens, E. R., Burleson, F. R., *J. Air Pollut. Control Assoc.* (1969) **19,** 929.
74. Stephens, E. R., Burleson, F. R., *J. Air Pollut. Control Assoc.* (1967) **17,** 147.
75. Daby, E. E., Stedman, D. H., Stuhl, F., Niki, H., "Measurement of Ozone and Nitrogen Oxides in Atmospheric and Laboratory Studies of Photochemical Smog," *Amer. Chem. Soc. Nat. Meet., 161st,* Los Angeles (March 1971) (in press, *J. Air Pollut. Control Assoc.*).
76. Donovan, R. J., Husain, D., Kusch, L. J., *Trans. Faraday Soc.* (1970) **66,** 2551.
77. Stuhl, F., Niki, H., *J. Chem. Phys.* (1971) **55,** 3943.
78. Calvert, J. G., Pitts, J. N., Jr., "Photochemistry," p. 368, Wiley, New York, 1966.
79. Daby, E. E., Stedman, D. H., Niki, H., "Mass Spectrometric Studies of the Reactions of Formaldehyde and Acetaldehyde with Atomic Oxygen in a Discharge Flow System," *Amer. Chem. Soc. Nat. Meet., 160th,* Chicago (September 1970).
80. Niki, H., *J. Chem. Phys.* (1966) **45,** 2330; (1967) **47,** 3102.
81. Cadle, R. D., Allen, E. R., "Chemical Reactions in Urban Atmospheres," p. 63, C. S. Tuesday, Ed., American Elsevier, New York, 1971.
82. Baulch, D. L., Drysdale, D. D., Lloyd, A. C., "High Temperature Reaction Rate Data," Department of Physical Chemistry, The University, Leeds, 2, England, No. 1 (1969).
83. Baulch, D. L., Drysdale, D. D., Lloyd, A. C., "High Temperature Reaction Rate Data," Department of Physical Chemistry, The University, Leeds, 2, England, No. 3 (1969).
84. Hoare, D. E., Pearson, G. S., *Advan. Photochem.* (1964) **3,** 83.
85. Niclause, M., Lemaire, J., Letort, M., *Advan. Photochem.* (1966) **4,** 25.
86. McMillan, G. R., Calvert, J. G., *Oxid. Combust. Rev.* (1965) **1,** 83.
87. Shtern, V. Ya., "The Gas-Phase Oxidation of Hydrocarbons," MacMillan, New York, 1964.
88. Hoare, D. E., Whytock, D. A., *Can. J. Chem.* (1967) **45,** 865, 2741, 2841.
89. Barnard, J. A., Cohen, A., *Trans. Faraday Soc.* (1968) **64,** 396.
90. Heicklen, J., *Advan. Chem. Ser.* (1968) **76,** 23.
91. Heicklen, J., *Int. Oxidation Symp.* (1967) **I,** 343.
92. Altshuller, A. P., Cohen, I. R., Purcell, T. C., *Can. J. Chem.* (1966) **44,** 2973.
93. Baldwin, R. R., Walker, R. W., *Trans. Faraday Soc.* (1969) **65,** 792.
94. Hoare, D. E., Patel, M., *Trans. Faraday Soc.* (1969) **65,** 1325.
95. Hay, J. M., Hessam, K., *Comb. Flame* (1971) **16,** 237.
96. Nicolet, M., *Ann. Geophys.* (1970) **26,** 531.
97. Nicolet, M., "Nitrogen Oxides in the Chemosphere," Scientific Report No. 227, The Pennsylvania State University, University Park, 1964.
98. Thomas, J. H., *Oxid. Combust. Rev.* (1965) **1,** 137.
99. Ashmore, P. G., Tyler, B. J., *Trans. Faraday Soc.* (1962) **58,** 1108.
100. Tyler, B. J., *Nature* (1962) **195,** 259.
101. Grätzel, M., Henglein, A., Taniguchi, S., *Ber. Bunsen-Gesell.* (1970) **74,** 292.
102. Nicholas, J. E., Norrish, R. G. W., *Proc. Roy. Soc.* (1969) **A309,** 171.

103. Phillips, L., Shaw, R., Tenth Symposium (International) on Combustion, p. 453 (1965).
104. Jaffe, L., Prosen, E. J., Szwarc, M., *J. Chem. Phys.* (1957) **27**, 416.
105. Vogt, T. C., Jr., Hamill, W. H., *J. Phys. Chem.* (1963) **67**, 292.
106. Greiner, N. R., *J. Chem. Phys.* (1970) **53**, 1284.
107. Wilson, W. E., "A Critical Review of the Gas Phase Reaction Kinetics of Several Bimolecular Reactions of the Hydroxyl Radical," NSRDS-NBS, in press.
108. Drysdale, D. D., Lloyd, A. C., *Oxid. Comb. Rev.* (1970) **4**, 157.
109. Berces, T., Trotman-Dickinson, A. F., *J. Chem. Soc.* (1961) **1961**, 4281.
110. Ashmore, P. G., Levitt, B. P., *Trans. Faraday Soc.* (1956) **52**, 835; (1957) **53**, 945.
111. Rosser, W. A., Wise, H., *J. Chem. Phys.* (1957) **26**, 571.
112. Berces, T., Forgeteg, S., *Trans. Faraday Soc.* (1970) **66**, 633, 640, 648.
113. Husain, D., Norrish, R. G. W., *Proc. Roy. Soc.* (1963) **A273**, 165.
114. McMillan, G. R., Kumari, J., Synder, D. L., "Chemical Reactions in Urban Atmospheres," p. 35, C. S. Tuesday, Ed., American Elsevier, New York, 1971.
115. Howard, J. A., Adamic, K., Ingold, K. U., *Can. J. Chem.* (1968) **46**, 2655, 2661; (1969) **47**, 3793, 3797, 3803 and earlier papers in this series.
116. Schott, G., Davidson, N., *J. Amer. Chem. Soc.* (1958) **80**, 1841.
117. Hisatsune, I. C., Zafonte, L., *J. Phys. Chem.* (1969) **73**, 2980.
118. Troe, J., *Ber. der Bunsen-Gesell.* (1969) **73**, 906.
119. Ford, H. W., Doyle, G. J., Endow, N., *J. Chem. Phys.* (1957) **26**, 1336.
120. Klein, F. S., Herron, J. T., *J. Chem. Phys.* (1964) **41**, 1285.
121. Altshuller, A. P., Cohen, I. R., *Int. J. Air Water Pollut.* (1963) **7**, 787.
122. Federal Register (1971) **36**, 8186.
123. Altshuller, A. P., Lonneman, W. A., Sutterfield, F. D., Kopczynski, S. L., *Environ. Sci. Technol.* (1971) **5**, 1009.

RECEIVED March 30, 1972.

3

Simulation of Urban Air Pollution

JOHN H. SEINFELD and STEVEN D. REYNOLDS

California Institute of Technology, Pasadena, Calif. 91109

PHILIP M. ROTH

Systems Applications, Inc., Beverly Hills, Calif. 90212

Several types of models are commonly used to describe the dispersion of atmospheric contaminants. Among these are the box, plume, and puff models. None are suitable, however, for describing the coupled transport and reaction phenomena that characterize atmospheres in which chemical reaction processes are important. Simulation models that have been proposed for the prediction of concentrations of photochemically formed pollutants in an urban airshed are reviewed here. The development of a generalized kinetic mechanism for photochemical smog suitable for inclusion in an urban airshed model, the treatment of emissions from automobiles, aircraft, power plants, and distributed sources, and the treatment of temporal and spatial variations of primary meteorological parameters are also discussed.

Urban airshed models are mathematical representations of atmospheric transport, dispersion, and chemical reaction processes which when combined with a source emissions model and inventory and pertinent meteorological data may be used to predict pollutant concentrations at any point in the airshed. Models capable of accurate prediction will be important aids in urban and regional planning. These models will be used for:

a. simulation of the effects of alternative air pollution control strategies on pollutant concentrations in the airshed

b. planning for land use so that projected freeways, industrial sites, and power plants may be located where their air pollution potential is minimized

c. determination of the long-term air pollution control strategy which accomplishes desired air quality objectives at least cost

d. real-time prediction in an alert warning system, such that an impending air pollution episode may be anticipated and proper preventive action taken.

Acceptance of air pollution models in decision-making will depend on the degree of confidence that can be placed in their predictions. The validity of a model is established by carefully comparing its predictions with air quality data for the particular airshed. Air quality models capable of predicting concentrations of primary (those emitted directly) and secondary (those formed by chemical reaction in the atmosphere) contaminants at any time and location in an airshed must clearly be based on sound fundamentals of meteorology and chemistry.

The particular type of air pollution model required depends on its expected uses. Models can be classified according to:

a. the spatial resolution of the predicted concentrations (the characteristic dimension over which major dependent variables such as wind and emissions are averaged)

b. the temporal resolution of the predicted concentrations (the characteristic time over which major dependent variables are averaged)

c. the complexity of the representation of advection, diffusion, and chemical reaction processes.

Typical horizontal spatial resolutions of a model are less than one square block, one square block to several square miles, or tens of square miles. The particular horizontal spatial scale chosen will depend on the size of the region being modeled and the purpose for the model. The smallest scale would be used, for example, to represent the canyon effects of tall buildings and the area surrounding a freeway. The intermediate scale is suitable for modeling an urban airshed. The largest scale would be applied to simulate pollutant transport over distances of regional to continental dimension. Vertical spatial resolution of a model is determined by the height where pollutants are expected to mix. Often vertical resolution is much finer than horizontal resolution because the vertical mixing depth in the airshed is considerably smaller than the horizontal distances where pollutants are advected. The Los Angeles Basin, for example, occupies roughly 2000 square miles, yet vertical pollutant transport is usually limited by a temperature inversion located at a height of about 1000–2000 feet.

Typical temporal resolutions of the predicted concentrations are several minutes to several hours and several months to one year. If the temporal resolution of the model is the order of several minutes to several hours, the model is generally used to compute concentrations over the course of one or more days. If the temporal resolution is several months to one year, the model is used to predict the average concentration in the airshed over such a time period, usually taking into account the frequency of occurrence of various meteorological conditions.

Several levels of complexity may be considered in the representation of advection, diffusion, and chemical reaction processes in the atmosphere. The simplest type of model is the so-called box model wherein pollutant concentrations are assumed to be homogeneous throughout the entire airshed. The airshed is thus treated as a giant well-mixed vessel (*1*). Also it is assumed that within the airshed: emitted pollutants are instantaneously and uniformly mixed, a uniform wind characterizes transport, and a constant inversion height is typical of time-averaged meteorology. A variant of the box model, consisting of a two-dimensional network of interconnected boxes or well-mixed cells, has been proposed by Reiquam (*2, 3, 4*). In this model the assumptions above apply to each cell although the cell volumes and the wind flows may change with time. Reiquam applied his model to estimate monthly average pollutant concentrations in the Willamette Valley and in Northern Europe. For horizontal spatial resolutions of tens of square miles and temporal resolutions of several months to one year, the box model may be used to estimate average concentration levels. However, when spatial resolutions of a few miles and temporal resolutions of the order of minutes are considered, the box model cannot be applied; it cannot properly describe variations in concentrations within the airshed.

The Gaussian plume and puff models, which describe the concentration distribution of an inert species downwind of a point, line, or area source, characterize the next level of complexity of airshed models. In the usual applications of these models:

1. Only inert pollutants are considered.

2. Wind shear is neglected.

3. Measures of plume and puff spread are based on experimental studies, are independent of height, and are a function of atmospheric stability class.

Plume models have been widely applied during the past decade to predict concentrations of CO, SO_2, and particulate matter in urban areas (*see*, for example (*5*), (*6*), (*7*), (*8*), (*9*), (*10*), (*11*), and (*12*)). These models cannot be applied, however, to predict pollutant concentrations under varying meteorological conditions or when chemical reactions are occurring. Nevertheless, Gaussian plume models often give useful estimates of concentrations downwind of strong isolated sources under steady meteorological conditions. The puff model, described by Roberts *et al.* (*13*), extends the plume model in that it is based on the Gaussian distribution and the same estimation procedures for dispersion parameters. However, by assuming that emissions can be treated as discrete puffs, certain assumptions normally made for the plume model are relaxed, notably that of steady state behavior. Chemical reactions cannot be included. Reviews of Gaussian plume and puff models have been presented by Lamb (*14*), Seinfeld (*15*), and Neiburger *et al.* (*16*).

The third level of complexity in airshed modeling involves the solution of the partial differential equations of conservation of mass. While the computational requirements for this class of models are much greater than for the box model or the plume and puff models, this approach permits the inclusion of chemical reactions, time-varying meteorological conditions, and complex source emissions patterns. However, since this model consists only of the conservation equations, variables associated with the momentum and energy equations—*e.g.*, wind fields and the vertical temperature structure—must be treated as inputs to the model. The solution of this class of models will be examined here.

Finally, the most complex and fundamental approach to air pollution modeling involves the solution of the full, three dimensional, time varying turbulent planetary boundary layer equations for conservation of mass, momentum, and energy. The computing speeds and storage capacities of the present generation of computers are not sufficiently great that the solution of such systems of equations is economically feasible. Thus, we will only briefly discuss this approach to atmospheric modeling.

A model based on the solution of the equations for the mean concentrations of each pollutant species requires several components:

1. a kinetic mechanism describing the rates of atmospheric chemical reactions as a function of the concentrations of the various species present

2. a source description, giving the temporal and spatial distribution of emissions from all significant pollutant sources in the airshed

3. a meteorological description, including wind speed and direction at each location in the airshed as a function of time, the vertical atmospheric temperature profile, and radiation intensity.

The overall model in which the components are imbedded is termed the core model. This core model along with the three components, comprises a comprehensive representation of the simultaneous transport, mixing, and chemical reaction processes which occur in the atmosphere.

Portions of the material described here are derived from a comprehensive airshed modeling program in which the authors are participating (*17*). This chapter focuses on urban airshed models; however novel models have been proposed for urban air pollution problems of a more restricted scale—particularly, the prediction of concentrations in the vicinity of major local sources, notably freeways, airports, power plants, and refineries. In discussing plume and puff models earlier we pointed out one such class of models. Other work is the model proposed by Eschenroeder (*18*) to predict concentrations of inert species in the vicinity of roadways and the modeling of chemically reacting plumes, based on the Lagrangian similarity hypothesis, as presented by Friedlander and Seinfeld (*19*).

The Governing Equations of Continuity

The equations of continuity form the basis for practical (*i.e.*, computable) urban airshed models. A short derivation of the semi-empirical equations of continuity (or the working equations) stating assumptions involved at each step, is presented followed by an expanded discussion of certain of these assumptions for a fuller understanding of the range of applications of the working equations and their shortcomings and limitations. While the material covered is largely a review of available knowledge, we believe that it is vital that individuals who develop or apply airshed models be clearly aware of the assumptions inherent in their derivation and thus of the restrictions that limit their applicability.

A Short Derivation of the Working Equations. Consider N species in a fluid. The concentration of each must at each instant in time satisfy a material balance over a volume element. Thus, any accumulation of material over time when added to the net amount of material convected into the volume element must be balanced by an equivalent amount of material that is produced by chemical reaction in the element, that is emitted into it, and that enters by diffusive transport. Expressed mathematically, the concentration of each species must satisfy the continuity equation,

$$\frac{\partial c_i}{\partial t} + \sum_{j=1}^{3} \frac{\partial}{\partial x_j} (u_j c_i) = \sum_{j=1}^{3} D_i \frac{\partial^2 c_i}{\partial x_j^2} + R_i (c_1, \ldots, c_N, T) + S_i (x,t) \quad (1)$$

$$i = 1, 2, \ldots, N$$

where u_j is the jth component of the wind velocity, D_i is the molecular diffusivity of the ith species in air, R_i is the rate of generation of species i by chemical reaction, and S_i is the addition rate of species i from volumetric sources. In addition to the requirement that the c_i satisfy Equation (1), the fluid velocities u_j and the temperature T, in turn, must satisfy the Navier–Stokes and energy equations, which are coupled through the u_j, c_i, and T with the total continuity equation and the ideal gas law.

Generally, it is necessary to carry out a simultaneous solution of the coupled equations of mass, momentum, and energy to account properly for the changes in T, c_i, and u_j and the effects of the changes in each of these variables on each other. In considering air pollution models, however, it is reasonable to assume that the presence of pollutants in the atmosphere does not affect the meteorology to any detectable extent; thus, the equations of continuity for contaminant species can be solved independently of the coupled momentum and energy equations. Nevertheless, despite this simplification, solution of the coupled momentum and energy equations is a substantial undertaking, the difficulties of which we discuss later. Most urban airshed models currently being developed are based solely on the equations of continuity for individual species. This restricted class of models is studied below, and upon this premise we pursue the derivation of the working equations, the semi-empirical equations of continuity.

Since atmospheric flows are turbulent, we may represent the wind velocities u_j as the sum of a deterministic and a stochastic component, $\hat{u}_j + u'_j$. The deterministic component includes that portion of the velocity which is known either from measurement or computation while the stochastic component represents everything not included in the deterministic component. Replacing u_j by $\hat{u}_j + u'_j$ in (1) gives

$$\frac{\partial c_i}{\partial t} + \sum_{j=1}^{3} \frac{\partial}{\partial x_j} \left[(\hat{u}_j + u'_j)c_i \right] = \sum_{j=1}^{3} D_i \frac{\partial^2 c_i}{\partial x_j^2} + R_i (c_1, \ldots, c_N, T) \qquad (2)$$
$$+ S_i (x, t)$$

Since the u'_j are random variables, the c_i resulting from the solution of (2) must also be random variables. (For convenience, we assume at this point that the temperature is constant and thus the dependence of R_i on T need not be explicitly indicated. It is not necessary to make this assumption, but it simplifies the presentation.)

Let us now assume that, since the c_i are random variables, each c_i can be decomposed into the sum of a deterministic and a stochastic component, $<c_i>$ and c'_i, respectively. Substituting into Equation (2) and averaging over all terms, we find that the mean concentration is governed by (assuming simple reactions)

$$\frac{\partial <c_i>}{\partial t} + \sum_{j=1}^{3} \frac{\partial}{\partial x_j} (\hat{u}_j <c_i>)$$

$$+ \sum_{j=1}^{3} \frac{\partial}{\partial x_j} <u'_j c'_i> = \sum_{j=1}^{3} D_i \frac{\partial^2 <c_i>}{\partial x_j^2} +$$

$$R_i (<c_1>, \ldots, <c_N>) + <R_i (c'_1, \ldots, c'_N)> + S_i(x, t) \qquad (3)$$

where we have assumed that $<c'_i> = 0$. We note the emergence of the cross product term $<u'_j c'_i>$ and the reaction rate $<R_i(c'_i, \ldots c'_N)>$, which includes terms of the form $<c'_i c'_k>$.

We now make two further assumptions, that molecular diffusion is negligible when compared with turbulent diffusion and that the atmosphere is incompressible. With reference to (3) this means that the first term on the right hand side of the equation is negligibly small compared with $\dfrac{\partial}{\partial x_j} <u'_j c'_i>$ and that the overall continuity equation $\overset{3}{\underset{j=1}{\Sigma}} \dfrac{\partial \hat{u}_j}{\partial x_j} = 0$ can be invoked. With these assumptions the species continuity equations become

$$\frac{\partial <c_i>}{\partial t} + \sum_{j=1}^{3} \hat{u}_j \frac{\partial <c_i>}{\partial x_j} + \sum_{j=1}^{3} \frac{\partial}{\partial x_j} <u'_j c'_i>$$

$$= R_i(<c_1>, \ldots, <c_N>) +$$

$$<R_i (c'_1, \ldots, c'_N)> + S_i (\underset{\sim}{x},t) \tag{4}$$

We now complete the derivation by introducing two final assumptions, that the turbulent flux $<u'_j c'_i>$ is linearly related to the gradients of $<c_i>$,

$$<u'_j c'_i> = - \sum_{k=1}^{3} K_{jk} \frac{\partial <c_i>}{\partial x_k} \qquad j = 1,2,3 \tag{5}$$

where K_{jk} is the (j,k) element of a turbulent eddy diffusivity tensor, which is generally an unknown function of position and time and that terms of the form $<c'_i c'_j>$ can be neglected. Invoking these two assumptions we find that

$$\frac{\partial <c_i>}{\partial t} + \sum_{j=1}^{3} \hat{u}_j \frac{\partial <c_i>}{\partial x_j} = \sum_{j=1}^{3} \sum_{k=1}^{3} \frac{\partial}{\partial x_j}\left(K_{jk} \frac{\partial <c_i>}{\partial x_k}\right)$$

$$+ R_i (<c_1>, \ldots, <c_n>) + S_i (\underset{\sim}{x},t) \tag{6}$$

If the coordinate axes coincide with the principal axes of the eddy diffusivity tensor K, only the three diagonal components of K are non-zero elements. Since there is no preferred horizontal direction we let these three components be $\{K_H, K_H, \text{and } K_V\}$. Equation (6) then reduces to

$$\frac{\partial <c_i>}{\partial t} + \sum_{j=1}^{3} \hat{u}_j \frac{\partial <c_i>}{\partial x_j}$$

$$= \frac{\partial}{\partial x}\left(K_H \frac{\partial <c_i>}{\partial x}\right) + \frac{\partial}{\partial y}\left(K_H \frac{\partial <c_i>}{\partial y}\right)$$

$$+ \frac{\partial}{\partial z}\left(K_V \frac{\partial <c_i>}{\partial z}\right) + R_i + S_i \tag{7}$$

This simplified form of (1) is the so-called semi-empirical equations of continuity, or, as we have referred to them, the working equations. It is the solution of these N coupled equations to which we referred earlier when we discussed the third level of complexity of airshed models. (The Gaussian plume model described above can be obtained by solving equation (7) under the assumptions that $\frac{\partial <c_i>}{\partial t} = 0$ and $\hat{u} = $ constant. The actual form of the plume model in current use is actually a modified version of the analytical solution, adopted for convenience of calculation.)

Discussion of the Working Equations. We have made numerous assumptions in deriving Equation (7) from the instantaneous material balance Equation (1). Several of these assumptions are related to actual atmospheric conditions and are generally valid in commonly occurring air pollution situations.

a. The presence of pollutants in the atmosphere does not cause variations to occur in temperature and velocity, and thus the equations of continuity can be solved independently of the equations of momentum and energy

b. Molecular diffusion is negligible

c. Atmospheric flow is incompressible

d. The system under examination is isothermal (an assumption of convenience only)

We have also made certain assumptions relating to the mathematical form of the equations:

e. Wind velocities and concentrations, as random variables, may be represented as the sum of deterministic and stochastic components with the average value of the stochastic components of concentration being zero (not really an assumption but a definition)

f. The turbulent fluxes are linearly related to the gradients in the mean concentration

g. Terms of the type $<c'_i c'_j>$, arising from the inclusion of chemical reactions, are negligible.

Here we examine assumptions (f) and (g), and we discuss the limitations inherent in the working equations (7) that restrict their applicability in describing the transport and reactions of air pollutants in the atmosphere.

The limitations associated with (7) are essentially a consequence of the stochastic nature of atmospheric transport and diffusion. Because the wind velocities are random functions of space and time, the airborne pollutant concentrations are random variables in space and time. Thus, the determination of the c_i, in the sense of being a specified quantity at any time, is not possible, but we can at best derive the probability density functions satisfied by the c_i. The complete specification of the probability density function for a stochastic process as complex as atmospheric diffusion is almost never possible. Instead, we must adopt a less desirable but more feasible approach, the determination of certain statisical moments of c_i, notably its mean, $<c_i>$. (The mean concentration can be interpreted in the following way. Suppose it were possible to repeat a particular day a large number of times with identical meteorological conditions. Since the wind velocities are turbulent, they would necessarily differ during each repetition. If we calculated an average of the pollutant concentrations experienced at each location and time over all the repetitions, we would have computed $<c_i(\underline{x},t)>$. Since days can not be repeated, a measurement of the concentration of species i is more suitably seen as one sample from the hypothetically infinite ensemble of identical days. Clearly, an individual measurement may differ considerably from the mean $<c_i>$.) We have done just this in our earlier derivation; Equation (4) is the equation satisfied by the mean concentrations. We now examine the nature of this equation and assumptions (f) and (g).

We note that Equations (4) contain dependent variables $<c_i>$ and $<u_j c_i>$, $j = 1,2,3$ for all i, and $<c'_i c'_j>$. We thus have more dependent variables than equations. In general, one possible means for eluding this problem is to generate equations which are actually continuity equations for the remaining dependent variables. For example, note that Equations (1) are written for c_i. We can derive a similar equation for the dependent variable $<u'_j c'_i>$ by subtracting (4) from (1), leaving an equation in c'_i. We then multiply this resulting equation by u'_j, and average over all terms. While we have derived the described equation, we have simultaneously generated new dependent variables $<u'^2_j c_i>$. In general, if we generate additional equations for the $<u'_j c'_i>$ and $<c'_i c'_j>$, we find that still more dependent variables appear in these equations. Following this procedure, we will always have more variables than equations. This problem, arising in the description of turbulent diffusion, is called the closure problem for which no general solution has yet been found.

To terminate the equations we must relate the product moments to those of lower order. The simplest, and most common approach for dealing with the $<u'_j c'_i>$ is that embodied in (5), the so-called eddy diffusivity assumption. No common methods exist for treating the product moments $<c'_i c'_j>$, arising when chemical reactions take place. It is assumed, therefore, that these terms are small compared with those of the type $<c_i><c_j>$. The key question is: Once we have invoked assumptions (f) and (g) as a means of closure of the equations, what limitations are inherent in (7) as a description of the mean concentrations of chemically reactive air pollutants?

This is not an easy question to answer, nor is it appropriate for us to address it in detail in a review paper. Rather, we refer to a detailed study of the validity of (7) as well as of other air pollution models that has been carried out by Lamb and Seinfeld (20). We restrict this discussion to a summary of the limitations in (7) which are introduced as a result of the eddy diffusivity hypothesis. In particular, Lamb and

Seinfeld have shown that the working equation (7) is valid, provided that the following conditions are met:

smooth over ranges of the order of the integration time step Δt and the distance over which a particle will travel in time interval Δt.

1. Spatial and temporal variations in the source emissions $S(\underline{x},t)$ are

2. The spatial and temporal inhomogeneities in the turbulence are of such scales that a particle released anywhere in the fluid will, over an interval Δt, behave as though it were in a field of stationary, homogeneous turbulence. In other words, the integration time step Δt should be much larger than the Lagrangian time scale of the turbulence.

3. The time scale of the rate controlling reactions described by $R(c_1, \ldots, c_N)$ is much larger than the Lagrangian time scale of the turbulence.

Assuming that these conditions are met and that the eddy diffusivities K_H and K_V are specified as functions of space and time, Equation (7) provides adequate representation of mean concentrations. Because the diffusivities are essentially empirical parameters to be determined from experimental data, the accuracy of (7) depends upon the degree to which atmospheric conditions at a location of interest correspond to the conditions under which the diffusivities were measured.

We have outlined those conditions under which the working equations apply; we must remember that (7) is *not* a universally valid set of equations for the mean concentrations of air pollutants. It can be justified only under rather limited circumstances. The problem of deriving more widely valid equations for atmospheric transport diffusion, and chemical reaction remains unsolved at this time.

Urban Airshed Models

Several approaches to airshed modeling based on the numerical solution of the semi-empirical equations of continuity (7) are now discussed. We stress that the solution of these equations yields the *mean* concentration of species i and not the actual concentration, which is a random variable. We emphasize the models capable of describing concentration changes in an urban airshed over time intervals of the order of a day although the basic approaches also apply to long time simulations on a regional or continental scale.

We divide the airshed models discussed here into two basic categories, moving cell models and fixed coordinate models. In the moving cell approach a hypothetical column of air, which may or may not be well mixed vertically, is followed through the airshed as it is advected by the wind. Pollutants are injected into the column at its base, and chemical reactions may take place within the column. In the fixed coordinate approach the airshed is divided into a three-dimensional grid, which can be envisioned as stacked layers of cells, each cell being perhaps 1–2 miles on a side and a few hundred feet high. This grid is then used as a basis for the numerical solution of the N coupled equations (7). In fluid mechanical terms the moving cell approach is Lagrangian, and the fixed coordinate approach is Eulerian. Each of the approaches has characteristics which suggest its use in particular air pollution modeling applications.

Moving Cell Approaches. As we have noted, the principal feature of the moving cell approach is that concentration changes in a hypothetical parcel of air are computed as the parcel traverses the airshed. The parcel is visualized as a vertical column of air of fixed area and variable height with the top of the column being defined by the base of an elevated inversion or, in the absence of an inversion, by an estimated maximum mixing height. The motion of the air column is assumed to correspond to the local instantaneous wind speed and direction, thereby tracing out a particular surface trajectory in the airshed.

The following assumptions are inherent in this model:

a. There is no horizontal transport of material across the boundaries of the parcel. (There is no way to include horizontal diffusive transport between the column and the environment.)

b. There is no change in the horizontal wind velocity with height.

c. Vertical advection is neglected—*i.e.*, the vertical component of the wind does not exist.

The basic assumption of the approach is that a parcel of air maintains its integrity while traversing the airshed. It is unlikely that this ever occurs in the atmosphere over the time scales of interest.

Since horizontal transport across the boundaries of the column is neglected and since it moves with the average ground-level horizontal wind velocity, the column may be mathematically represented as a horizontally uniform but vertically non-uniform column with a time-varying source of pollutants at the base. Thus, the only independent variables are time t and vertical distance z. The concentration of species i at time t and height z in the column $c_i(z,t)$ (From this point on we use c_i to denote $<c_i>$, the mean concentration of species i.) is determined by integration of the abridged form of (7),

$$\frac{\partial c_i}{\partial t} = \frac{\partial}{\partial z}\left(K_V(z)\frac{\partial c_i}{\partial z}\right) + R_i(c_1, \ldots, c_n) + S_i \tag{8}$$

The initial condition for (8) is that the concentration within the column at the beginning of the traversal be given—*i.e.*,

$$c_i(z,0) = c_{i_o} \tag{9}$$

The boundary conditions on z at the ground, $z = 0$, and the inversion base (or top of the column), $z = H(t)$, are given by

$$- K_V(0) \frac{\partial c_i}{\partial z} = Q_i(t) \quad z = 0 \tag{10}$$

$$- K_V(H) \frac{\partial c_i}{\partial z} = 0 \quad\quad z = H(t) \tag{11}$$

where $Q_i(t)$ is the flux of species i from ground-level sources, S_i is the volumetric rate of emission of species i from elevated sources, and $H(t)$ is the height of the column as a function of time. The movement of the column is mathematically reflected only in $Q_i(t)$ and $H(t)$. An approach based on (8) to (11) has been developed by Eschenroeder and Martinez (*21*) for Los Angeles.

A simplified version of (8) results if we neglect vertical inhomogeneities in the column. Then $c_i = c_i(t)$ only and (8) reduces to

$$\frac{d (Vc_i)}{dt} = V R_i (c_1, \ldots ,c_n) + AQ_i + VS_i \tag{12}$$

where the column is simply a well-mixed vessel having a volume V, a base area A, and a time varying pollutant input $AQ_i + VS_i$. The advantage of using (12) over (8)–(11) is mainly computational since (12) consists only of a set of ordinary differential equations rather than a set of partial differential equations. Wayne *et al.* (*22*) have used an approach based on (12) for the Los Angeles Basin.

We stress that the moving-cell approach is *not* a full airshed model nor is it intended as such. Rather, it is a technique for computing concentration histories along a given air trajectory. It is not feasible to use this approach to predict concentrations as a function of time and location throughout an airshed since a large number of trajectory calculations would be required.

The moving cell approach has the following advantages:

a. It is computationally advantageous to avoid the integration of (7) in all three spatial dimensions and time.

b. The concentration history along an air trajectory can be traced, thereby permitting an assessment of the effect of specific sources at locations downwind of these sources.

It has the following deficiencies:

a. The concept of an identifiable parcel of air is oversimplified since such an entity never exists in a turbulent atmosphere over time scales of interest.

b. There is no way to include convergence and divergence phenomena in the wind field, and the resulting vertical advection of air.

c. To determine the concentration at a given location and time, it is necessary to trace the trajectory backward in time to the point where it entered the airshed. Since the only reason for this calculation is to ascertain the starting point of the trajectory, its inclusion constitutes an inefficiency inherent in the approach, particularly when a large number of trajectories must be computed.

Fixed Coordinate Approaches. In the fixed coordinate approach to airshed modeling, the airshed is divided into a three-dimensional grid for the numerical solution of some form of (7), the specific form depending upon the simplifying assumptions made. We classify the general methods for solution of the continuity equations by conventional finite difference methods, particle in cell methods, and variational methods. Finite difference methods and particle in cell methods are discussed here. Variational methods involve assuming the form of the concentration distribution, usually in terms of an expansion of known functions, and evaluating coefficients in the expansion. There is currently active interest in the application of these techniques (23); however, they are not yet sufficiently well developed that they may be applied to the solution of three-dimensional time-dependent partial differential equations, such as (7). For this reason we will not discuss these methods here.

The principal numerical problem associated with the solution of (7) is that lengthy calculations are required to integrate several coupled nonlinear equations in three dimensions. However, models based on a fixed coordinate approach may be used to predict pollutant concentrations at all points of interest in the airshed at any time. This is in contrast to moving cell methods, wherein predictions are confined to the paths along which concentration histories are computed.

FINITE-DIFFERENCE METHODS. The numerical analysis literature abounds with finite difference methods for the numerical solution of partial differential equations. While these methods have been successfully applied in the solution of two-dimensional problems in fluid mechanics and diffusion (24, 25), there is little reported experience in the solution of three-dimensional, time-dependent, nonlinear problems. Application of these techniques, then, must proceed by extending methods successfully applied in two-dimensional formulations to the more complex problem of solving (7). The various types of finite-difference methods applicable in the solution of partial differential equations and their advantages and disadvantages are discussed by von Rosenberg (26), Forsythe and Wasow (27), and Ames (28).

The principal considerations in choosing a finite-difference method for (7) are accuracy, stability, computation time, and computer storage requirements. Accuracy of a method refers to the degree to which the numerically computed temporal and spatial derivatives approximate the true derivatives. Stability considerations place restrictions on the maximum time step Δt that can be used in the integration. Implicit methods, those involving the simultaneous solution of difference equations at each step, are more suitable for solving nonlinear forms of (7) than are explicit methods, as the former are stable over a wider range of step sizes. Implicit methods, however, involve considerably more computation per time step than do explicit methods. Other finite difference methods exist which are difficult to classify. Typically, these techniques have the characteristics of implicit methods but because of some unique aspects of the particular method involve less burdensome calculations than are normally expected with an implicit method. Two such techniques that have the potential for application in the solution of (7) are fractional steps method (29) and the alternating directions method (30).

There have thus far been reported only two applications of finite-difference methods to the solution of (7) as they pertain to urban airsheds, both for the Los Angeles Basin. Eschenroeder and Martinez (21) applied the Crank–Nicolson implicit method to the simplified version of, (7),

$$\frac{\partial c_i}{\partial t} + \hat{u} \frac{\partial c_i}{\partial x} = \frac{\partial}{\partial z}\left(K_V(z) \frac{\partial c_i}{\partial z}\right) + R_i$$

In a later paper (31) they report that a number of difficulties were encountered using the Crank–Nicolson method and the approach was abandoned. More recently, Roth *et al.* (17) have applied the method of fractional steps to the solution of six equations of the form of (7), four of which are coupled; this effort is continuing.

The main advantage in using a finite-difference method to solve (7), as compared with other approaches, is that there has been extensive experience in applying these techniques to various partial differential equations. Even though reported experience with three-dimensional, time-dependent, nonlinear problems is sparse, experience with simpler systems gives a sound basis to develop feasible approaches. The disadvantages of finite difference methods are well-known:

a. Inaccuracies in approximating the first-order advection terms in the continuity equations give rise to second-order errors, which have the mathematical characteristics of diffusion processes. These inaccuracies, often termed numerical or artificial diffusion, can mask the representation of true diffusion. Finite difference methods based on approximations higher than second order are required to minimize this difficulty; computational times increase with increasing order of the method used.

b. Computing time and storage requirements associated with accurate, stable methods can be substantial for problems involving several independent variables. When the equations are nonlinear, time-consuming iterations or matrix inversions are often required in the solution.

PARTICLE IN CELL METHODS. An alternative to the direct finite-difference solution of (7) is the so-called particle in cell (PIC) technique. The distinguishing feature of the PIC technique is that the continuous concentration field is treated as a collection of mass points, each representing a given amount of pollutant and each located at the center of mass of the volume of material it represents. The mass points, or particles, are moved by advection and diffusion. It is convenient but not necessary, to have each of the particles of a given contaminant represent the same mass of material. The application of the PIC technique in hydrodynamic calculations is discussed by Harlow (32). Here we consider the use of the PIC technique in the numerical solution of (7).

Given an initial, continuous concentration field in the airshed, we replace this field by discrete particles of pollutant i, each representing a fixed mass. The particles are located within a three dimensional, fixed grid according to the mass distribution of material. Thus, each particle has a given set of coordinates. Consider now a single time step Δt in the numerical solution of (7), using the PIC method. We write (7) in the form

$$\frac{\partial c_i}{\partial t} + \nabla \cdot \underline{U}_i c_i = R_i + S_i \tag{13}$$

where the effective velocity U_i is defined by

$$\underline{U}_i = \hat{\underline{u}} - \frac{K}{c_i} \nabla c_i \tag{14}$$

and $K = [K_H, K_H, K_V]$. In the computational procedure each particle of species i at location (x,y,z) is moved a distance $|U_i \Delta t|$ in the direction of U_i over a time step Δt. Also new particles are emitted during the period t to $t + \Delta t$ from the sources located in each cell, the number of particles emitted being determined by the product of the source strength and the time step. These new particles are also convected with velocity $\underline{U}_i$. After the convective step the average concentration of each species in a cell is calculated, this concentration being equal to the total mass of particles occupying the cell, divided by the cell volume. The cell contents are then allowed to react, resulting in a concentration change, $R_i \Delta t$. Finally, particles are reconstituted with the change in mass owing to chemical reaction being reflected in changes in the number of particles of each species. The same procedure is then repeated for succeeding time steps. The PIC technique has been adapted to air pollution modeling by Sklarew (33).

The PIC technique has the following advantages:

a. Artificial diffusion because of truncation errors in the advection terms in (7) is eliminated since these terms are not approximated by finite-difference representations

b. There are no stability restrictions on Δt (although Δt should be small enough so that the value of U_i represents the movement of fluid particles)

c. Particles can be tagged as to their place of origin, thus making it possible to identify the sources of contaminants observed at any location and the following weaknesses:

1. Computer storage requirements can become excessive, as the co-ordinates of a large number of particles must be kept in memory

2. If it is assumed that each particle of a given contaminant represents the same mass of material, then every cell will have a residue that cannot be assigned to a particle. On the average this residual material will equal one-half of a particle mass. If the assumption that particles be of equal mass is relaxed, the residue error can be eliminated but only at the cost of storing a large amount of additional information—the masses of all species in each cell.

WELL-MIXED CELL MODEL. A conceptually simple approach is based on the representation of the airshed by a three-dimensional array of well-mixed vessels (*34, 35, 36*). As before, we assume that the airshed has been divided into an array of L cells. Instead of using the array simply as a tool in the finite-difference solution of the continuity equations, let us now assume that each of these cells is actually a well-mixed reactor with inflows and outflows between adjacent cells. If we neglect diffusive transport across the boundaries of the cells and consider only convective transport among cells, a mass balance on species i in cell k is given by

$$v_k \frac{dc_{ik}}{dt} = -c_{ik} \frac{dv_k}{dt} + \sum_{j=0}^{L} q_{jk}c_{ij} \tag{15}$$

$$-c_{ik} \sum_{j=0}^{L} q_{kj} + S_{ik} + R_i(c_{1k}, \ldots, c_{nk})$$

where

c_{ik} = concentration of species i in cell k
v_k = volume of cell k
q_{jk} = volumetric flow rate of air from cell j to cell k, q_{ok} is the flow from the exterior of the airshed into cell k, and q_{ko} is the flow from cell k to the exterior of the airshed
S_{ik} = rate of emission of species i into cell k from all sources
R_i = rate of formation of species i by chemical reaction

Normally dv_k/dt is set equal to $A_k(dH_k/dt)$, where A_k is the area of the base of a cell having vertical sides and H_k is the height of the top of the cell. Actually the cell is a box with permeable walls and a movable lid.

If we divide the airshed into L cells and consider N species, LN ordinary differential equations of the form (15) constitute the airshed model. As might be expected, this model bears a direct relation to the partial differential equations of conservation (7). If we allow the cell size to become small, it can be shown that (15) is the same as the first-order spatial finite difference representation of (7) in which turbulent diffusive transport is neglected—i.e.,

$$\frac{\partial c_i}{\partial t} + \hat{u} \cdot \nabla c_i = R_i + S_i$$

Therefore, the well-mixed cell model can also be described as the result of the finite difference approximation of the spatial derivatives of (7)— i.e., of the conservation equations in which diffusion has been neglected.

The advantages of the well-mixed cell approach are as follows:

a. The geometries of cell bases (which may be irregular and variable from cell to cell) can be drawn to conform with topographic features.

b. Variations in inversion height with time are easily incorporated in the model.

c. The model is conceptually easy to understand and implement (only ordinary differential equations are involved).

Its disadvantages, however, are considerable:

a. Resulting from the large variations that can occur in the magnitudes of the flows, q_{jk}, equations (15) are are often stiff, thus requiring implicit integration techniques to insure stability in their solution. If an implicit technique is used, the inversion of an $NL \times NL$ matrix is necessary at each time step. Since computing the inverse of large matrices can be time-consuming, this requirement places a definite restriction on the size of L. For example, if we were to consider 25 cells for our system of four coupled equations, the repeated inversion of a 100×100 matrix would be required.

b. Diffusive transport is neglected. This is a distinct drawback for vertical diffusion.

c. The mathematical formation of the well-mixed cell model, Equations (15), is such that the expected accuracy of the solution is equivalent to that expected from the application of only a first order finite difference method to the solution of a corresponding model based on the partial differential equation, (7).

MacCracken et al. (36) have applied the well-mixed cell model in describing pollutant transport and dispersion in the San Francisco Bay Area.

Kinetics of Atmospheric Chemical Reactions

Chemical reaction processes account for the production of a variety of contaminant species in the atmosphere. Each of the basic airshed models above includes reaction phenomena in the conservative equations. The reaction term, denoted by R_i, accounts for the rate of production of species i by chemical reaction and depends generally on the concentrations of each N species. The conservation equations are thus coupled through the R_i terms, the functional form of each term being determined through the specification of a particular kinetic mechanism for the atmospheric reactions.

There will be instances where the use of an airshed model will be limited to the prediction of concentrations of inert species. However, when chemical reaction processes are important, it is essential to include an adequate description of these phenomena in the model. Here we outline the requirements that an appropriate kinetic mechanism must meet, survey pertinent model development efforts, and present an example of a mechanism that possesses many of the attributes that a suitable model must display.

General Considerations. The nature and characteristics of atmospheric contaminants suggest certain difficulties in the formulation of a kinetic mechanism of general validity. First, there is a multiplicity of stable chemical species in the atmosphere. Most species are present at low concentrations, thereby creating major problems in detection and analysis. A number of atmospheric constituents probably remain unidentified. Also, there are a large number of short-lived intermediate species and free radicals which participate in many individual chemical reactions. However, while we must admit to only a partial understanding of atmospheric reaction processes, it remains essential that we attempt to formulate quantitative descriptions of these processes which are suitable for inclusion in an overall simulation model.

A suitable mechanism must not be overly complex since computation times for the integration of the basic model in which the mechanism is to be imbedded are likely to be excessive. However, too simplified a mechanism may omit important reaction features. A major issue in this regard is that the mechanism predict the behavior of a complex mixture of many hydrocarbons, yet do so with minimum detail. The goal is to achieve acceptable accuracy in prediction without an undue computational burden.

Kinetic mechanisms that have been proposed fall into two general categories, detailed mechanisms for photooxidation of a single hydrocarbon and compact, generalized mechanisms for a complex mixture. Detailed mechanisms which attempt to account for the history of all species generated must be ruled out for three reasons. First, while the aim of those developed thus far has been completeness of description, this thoroughness has been achieved by including numerous reaction steps that involve free radicals. Knowledge of the rates of these reactions is imprecise. Furthermore, when several free radical reactions are included in a mechanism, the flexibility in the choice of rate constants is increased as each imprecisely known parameter can be varied independently in the process of matching prediction and experiment. To the extent that detailed mechanisms possess this flexibility in parameterization, the validity of comparison of prediction and experiment is diminished. Second, computation time is a limiting factor in the solution of the coupled partial differential equations that comprise the overall airshed model. The inclusion of a detailed mechanism in such a model greatly increases the computational burden and is to be avoided if possible. Finally, the decision to develop and implement a detailed mechanism implies the desire to represent reaction processes as accurately as is feasible. Thus, a relatively large number of reaction steps must be incorporated in a description of the dynamics of consumption of a particular hydrocarbon, such as propylene. Reaction dynamics will, however, vary for the many hydrocarbon species present in the atmosphere. If, for example, 30 to 40 steps are required to describe propylene kinetics, and 50 hydrocarbon species, each having unique dynamics, are believed to exert a significant impact on atmospheric reaction processes, one is faced with an intractable representation of the system.

The kinetic mechanism, once developed, must be validated. This process is commonly thought to consist of two parts, validation in the absence of transport-limiting steps and validation in their presence. In practical terms we are speaking, respectively, of comparing the model's predictions with data collected in smog chamber experiments and with data collected at actual monitoring stations situated in an urban airshed. When we speak of validation of a kinetic mechanism in this section, we are referring to the comparison between predictions and experiment based on smog chamber studies. The second and more complex of the two parts, validation of the kinetic mechanism in the presence of transport-limiting steps, is the primary undertaking of the overall modeling effort.

Only the most intrepid readers can have been exposed to the warnings and qualifications that highlight this discussion and can have emerged with undiminished faith, yet one final warning is necessary. Smog chamber experiments may be considered valid simulations of the atmosphere only under the following conditions: (1) wall effects are eliminated or are negligible, (2) contaminant concentrations are at levels found in the atmosphere, (3) the initial charge to the chamber is representative of urban source effluents, and (4) the spectral distribution of the radiation is the same as that found in the atmosphere. It is unlikely that many smog chamber studies satisfy all these requirements.

To summarize, then, we require a mechanism which: (1) describes reaction rate phenomena accurately over a specified range of concentrations, (2) is a parsimonious representation of the actual atmospheric chemistry, in the interest of minimizing computation time, and (3) can be written for a general hydrocarbon species, with the inclusion of variable stoichiometric coefficients to permit simulation of the behavior of the complex hydrocarbon mixture that actually exists in the atmosphere. Thus, we seek a mechanism which incorporates a balance between accuracy of prediction and ease of computation.

A Kinetic Mechanism. Relatively little work has been carried out until quite recently in the formulation of kinetic mechanisms for atmospheric reactions. (This is in contrast to the considerable efforts that have been expended by numerous investigators over the past two decades in the study of individual atmospheric reactions. Since the general literature in the field of atmospheric chemistry is voluminous, the reviews of Leighton (37), Altshuller and Bufalini (38, 39), and Johnston *et al.* (40) are recommended.) The first systematic study is apparently that of Wayne (41). More recent efforts have been reported by Eschenroeder (42), Behar (43), Westberg and Cohen (44), Wayne *et al.* (23), Hecht and Seinfeld (45), and Eschenroeder and Martinez (32). Of these the earlier stuides are typified by the development of mechanisms (19, 42, 43) which reproduced the gross features of the photochemical smog system but showed deficiencies in representing the effects of changes in initial reactant concentrations while neglecting altogether the effects of CO and H_2O on the system. More recently, improved mechanisms have been proposed; two of the most promising are those of Hecht and Seinfeld (45) and Eschenroeder and Martinez (32).

An example of a generalized mechanism suitable for inclusion in an urban airshed model is presented below. In particular, we wish to illustrate the scope of such a mechanism and the level of detail that must be included to ensure accuracy while avoiding undue complexity. We have selected the mechanism proposed by Hecht and Seinfeld (45) for this purpose. This mechanism fulfills the requirements for suitability summarized earlier, has predicted accurately the concentration–time behavior of pollutant species in a smog chamber for a variety of hydrocarbon–NO_x mixtures, and can be included in any of the airshed models described without difficulty. We note, however, that the mechanism is not a unique description of atmospheric chemistry; modified and improved versions may well be developed during the next few years.

Table I. A Kinetic Mechanism for Photochemical Smog

$$NO_2 + h\nu \xrightarrow{\ 1\ } NO + O$$

$$O + O_2 + M \xrightarrow{\ 2\ } O_3 + M$$

$$O_3 + NO \xrightarrow{\ 3\ } NO_2 + O_2$$

$$O_3 + 2NO_2 \xrightarrow[H_2O]{\ 4\ } 2HNO_3 + O_2 \; ^a$$

$$NO + NO_2 \xrightarrow[H_2O]{\ 5\ } 2HNO_2$$

$$HNO_2 + h\nu \xrightarrow{\ 6\ } OH\cdot + NO$$

$$CO + OH\cdot \xrightarrow[O_2]{\ 7\ } CO_2 + HO_2\cdot \; ^b$$

$$HO_2\cdot + NO_2 \xrightarrow{\ 8\ } HNO_2 + O_2$$

$$HC + O \xrightarrow{\ 9\ } \alpha RO_2\cdot$$

$$HC + O_3 \xrightarrow{\ 10\ } \beta RO_2\cdot + \gamma RCHO$$

$$HC + OH\cdot \xrightarrow{\ 11\ } \delta RO_2\cdot + \varepsilon RCHO$$

$$RO_2\cdot + NO \xrightarrow{\ 12\ } NO_2 + \theta OH\cdot$$

$$RO_2\cdot + NO_2 \xrightarrow{\ 13\ } \text{PRODUCTS (INCL. PAN)}$$

$$HO_2\cdot + NO \xrightarrow{\ 14\ } NO_2 + OH\cdot$$

[a] Reaction 4 is a combination of the three reactions:

$$O_3 + NO_2 \xrightarrow{\ 4a\ } NO_3 + O_2$$

$$NO_3 + NO_2 \xrightarrow{\ 4b\ } N_2O_5$$

$$N_2O_5 + H_2O \xrightarrow{\ 4c\ } 2HNO_3$$

[b] Reaction 7 is a combination of the two reactions:

$$CO + OH\cdot \xrightarrow{\ 7a\ } CO_2 + H\cdot$$

$$H\cdot + O_2 \xrightarrow{\ 7b\ } HO_2\cdot$$

Table I gives a complete statement of the mechanism; reference should be made to this table throughout the following discussion. The major inorganic species that participate in photochemical smog reactions are NO, NO_2, O_2, CO, O_3, and H_2O. In Table I Reactions 1–8 and 14 represent most of the important reactions among these species and account for the following experimentally observed phenomena:

a. The primary inorganic reactions in the system of NO, NO_2, O_3 (Reactions 1, 2, and 3)

b. Formation of nitric acid (Reaction 4)

c. Formation and photolysis of nitrous acid (Reactions 5, 6, and 8)

d. The reaction of CO and OH· radicals (Reaction 7).

Reactions 5 and 6 have been included to account for the importance of HNO_2 as a source of OH· radicals in the presence of water. In a dry system Reactions 4 and 5 will be omitted. Reaction 8 is included to provide for the consumption of HO_2· when NO has been depleted.

Turning now to the hydrocarbon reactions, we introduce the species HC to represent a hydrocarbon mixture or an individual hydrocarbon for a smog chamber experiment. The important hydrocarbon oxidation reactions are those with atomic oxygen, ozone, and hydroxyl radicals. Rather than attempt to identify the specific products of these individual reactions, we introduce the lumped radical species RO_2· as representing the total population of oxygen-containing free radicals which result from the three hydrocarbon oxidation reactions. Under these assumptions, the hydrocarbon reactions may be represented by three reaction steps, given by Reactions 9, 10, and 11. (*See* Hecht and Seinfeld (*45*) for a description of the rationale of this formulation.) We note that the coefficients α, β, and δ represent the number of RO_2· radicals formed in each reaction, that their magnitudes depend on the particular hydrocarbon (or mixture), and that they must be established empirically.

The remaining organic reactions in the simplified mechanism, 12 and 13, describe the oxidation of NO and NO_2 by peroxy radicals and the formation of PAN. We emphasize that if CO and H_2O are present, HO_2· and RO_2· are treated as separate species. In the absence of CO and H_2O, however, only the single species RO_2·, which includes HO_2· within it, is considered since Reactions 4–8 and 14 are omitted. In this cases θ (Reaction 12) represents the fractional product of OH· in the total lumped radical species RO_2·.

The mechanism, then, consists of Reactions 1–14 in Table I and includes the following species: NO, NO_2, O_3, HC, O, OH·, HO_2·, RO_2·, HNO_2, HNO_3, RCHO, and PAN. Differential equations are required for the first four species, steady state relations for the next five. The last three species are products and may also be represented by differential equations. (Since CO and H_2O are generally present in high concentrations, their concentrations may be assumed constant.) For example, in considering an airshed model formulated to describe the concentration time behavior of hydrocarbons, nitrogen oxides, oxidants, and carbon monoxide, it is only necessary to substitute into the R_i term of the continuity equation for each of these four species the appropriate differential equation describing the reaction behavior of that species.

Validation of the Mechanism. The process of matching the predictions of the mechanism to experimental smog chamber data is termed validation of the mechanism. The first step in a validation procedure is to establish values for the two major classes of parameters that appear in the mechanism—the reaction rate constants and the stoichiometric coefficients. Base values of the rate constants can be estimated from the chemical literature. However, with the sacrifice of chemical detail present in the new, simplified mechanism is a loss in the ability to associate the rate constant values with particular reactions. Therefore, the rate constants in the simplified mechanism are more a quantitative assessment of the relative rates of competing reactions than a reflection of the exact values for particular reactions. Base values for the parameters that appear in the kinetic mechanism are thus established on the basis of published rate constants. However, we must expect that final validation values will consist of those values which produce the best fit of the mechanism to actual smog chamber data. A recent summary of rate constants for specific hydrocarbon systems was made by Johnston *et al.* (*40*) from which rate constants for the Reactions in Table I can be estimated for a number of hydrocarbons.

As was pointed out above, a problem arises in the specification of the stoichimetric coefficients which appear in the hydrocarbon reactions in the mechanism ($\alpha, \beta, \delta, \epsilon, \theta$). For well-defined reaction steps such as the inorganic reactions, coefficients may be calculated through a simple mass balance. However, generalized stoichimetric coefficients which appear in the hydrocarbon reactions must first be estimated through deductive procedures and then established during validation. Their values will generally depend on the characteristics and composition of the system under study.

The mechanism described has been the subject of numerous validation exercises. The smog chamber experiments against which the mechanism has been tested include the following hydrocarbons: propylene, isobutylene in the presence and absence of CO, *n*-butane in the presence and absence of CO, and a mixture of propylene and *n*-butane. Predicted concentrations generally match well with experimental results for all systems studied. (*See* Hecht and Seinfeld (*45*) for a detailed description of the validation procedure and results.)

While a validated photochemical mechanism may closely represent the concentration/time behavior of smog chamber experiments, two factors must be considered before the mechanism can become part of an urban airshed model. These are variations in reactivity of the atmospheric hydrocarbon mixture and variations in radiation intensity, which influence the rates of decomposition of NO_2 and HNO_2 (Reactions 1 and 6). Reactivity variations may be classified as variations in composition, in location within the airshed, and in time. Eschenroeder and Martinez (46) in a detailed study of Los Angeles air quality data were unable to detect variations in reactivity with location or time. However, such variations may be observed and should be properly accounted for. Compositional variations may be included in an airshed model by including two or more generalized hydrocarbon species. In the simplest case, two species, one might consider a reactive and an unreactive hydrocarbon. Reactivity of individual atmospheric hydrocarbon would be assigned according to a scale of reactivity, such as that of Altshuller (47) or Bonamassa and Wong-Woo (48). The reactive species would be equivalent to HC in the reaction mechanism while the unreactive would be treated as inert. It may be necessary, however, to include more than two categories of hydrocarbon to represent the reactivity of the atmospheric mixture with sufficient precision.

Many variables influence the intensity and spectral distribution of radiation that reaches the lower atmosphere; Leighton (37) thoroughly discusses these factors. It is not yet possible to incorporate quantitatively into an airshed model the effects of many of these variables on radiation characteristics. In a simple treatment, Seinfeld et al. (49) have considered the variations in radiation intensity with latitude (for Los Angeles) with time of year and time of day. Wayne et al. (22) have adopted portions of the model suggested by Leighton. Much more work is needed, however, before the influence of incoming radiation can be properly included in airshed models.

Source Emissions

Perhaps the most tedious and mundane aspect in the development and validation of an atmospheric simulation model is the compilation of a complete contaminant emissions inventory. Yet, such an inventory must be made before a model can be validated since the spatial and temporal distribution of contaminant emissions comprises a direct input to the overall simulation model. Ground-level sources enter into the boundary conditions of the conservation equations through the function $Q_i(x,y,t)$ introduced previously; elevated sources enter as $S_i(x,y,z,t)$ in the conservation equations themselves.

The major sources of pollutant emissions may be classified as moving and fixed. The predominant moving source in all urban airsheds is vehicular traffic, primarily automobiles and trucks. The airplane is a much less significant contributor, and all other moving sources are usually neglected. Whereas the contributions of aircraft to total contaminant emissions in an urban airshed may be small, the percentages of pollutants near airports that are attributable to aircraft operations are often significant. Contaminants emitted from moving sources consist mainly of carbon monoxide, hydrocarbons, nitrogen oxides, and particulates.

Power plants and refineries are the primary fixed sources of pollutant emissions in most urban areas. However, other industrial sources, distributed throughout the area, also emit substantial amounts of contaminants. Also during the winter months effluents from home heating can add significantly to the pollutant load in the atmosphere. Sulfur dioxide and particulates are emitted from nearly all fixed sources although many of the particulate emissions are controlled by the use of abatement devices and sulfur dioxide by the use of low sulfur fuels. Also, power plants emit large amounts of nitrogen oxides, and refineries, hydrocarbons.

There is a multiplicity of models for pollutant emissions that may be applied to individual sources and source types. The model that is used and the degree of detail that is incorporated depend upon the spatial and temporal resolution of the overall airshed model, the type and amount of available data, and the accuracy of those data. For example, in attempting to estimate contours of pollution concentrations over the Los Angeles Basin during the course of a day and under particular meteorological conditions, it is necessary to compute the distribution over space of pollutant emissions from automobiles with a resolution of the order of one mile, and over time with a resolution of the order of one hour. Detailed traffic count data are available in nearly all parts of the Basin; whereas, average vehicle speed over various routes and variations in this quantity with time are available only for freeways. With such data, we can account for temporal variations in emissions rates from an average vehicle on freeways. However, we are unable to account for similar variations in average speed on main arteries, owing to a lack of such data. In other cities it may well be that emissions/average speed data are unavailable for any roadways. In such circumstances one may either adopt mean emissions factors or assume a temporal distribution of average speeds. There is more than one way to describe quantitatively the emissions from a particular type of pollutant source. The choice of approach depends on factors specific to the modeling effort and the goals of the project and on the availability of data.

Examples of models applicable to emissions from automotive sources, aircraft, and power plants are presented below. Distributed sources may be treated as fluxes included in the boundary conditions or in special cases as volume sources. Most of the models to be described have been used by the authors to develop an airshed model for the Los Angeles Basin (50). This particular application involved a two-mile spatial resolution and a one hour temporal resolution.

Automotive Emissions. The magnitude of contaminant emissions from a motor vehicle is a variable in time and is a function of the percentage of time the vehicle is operated in each driving mode—accelerate, cruise, decelerate, idle. Also affecting emissions are the presence or absence of a smog control device, the car's condition, its size, and other factors. The distribution of vehicles in time and space throughout an urban area is similarly governed by a number of factors—*e.g.*, commuting routes, distribution of centers of employment, and working hours. Precise calculation of the magnitude of motor vehicle emissions as a function of location and time is clearly not possible.

While all pertinent factors that affect the distribution of vehicular emissions cannot be taken into account, emissions rates may still be represented with sufficient accuracy to merit inclusion in an urban airshed model. For simplicity a vehicle emissions inventory can be divided into two parts:

a. estimation of spatial and temporal distribution of traffic

b. estimation of average vehicle emissions rates applicable to traffic in the area

The spatial and temporal distribution of traffic on the freeways and surface streets in an urban area can be estimated from traffic counts which are normally taken by state and local agencies. Vehicle exhaust emissions rates are estimated from data collected in tests that simulate the emissions of vehicles actually driven over typical routes in the urban area being studied. The California driving cycle and the 1972 Federal test procedure are two such tests. The flux of contaminants from motor vehicles into a grid square is conveniently computed with the necessary data as $Q_k M_l$, where Q_k is the mass of species k emitted per vehicle mile from an average vehicle, and M_l represents the vehicle miles travelled in the grid square during hour l.

Average pollutant emissions rates must be computed in a different manner for each of the three types of automotive emissions—exhaust, crankcase leakage, and evaporative losses. For an uncontrolled vehicle:

a. Exhaust emissions account for about 65% of hydrocarbons and 100% of nitrogen oxides and CO

b. Crankcase leakage (or blow-by) accounts for about 20% of hydrocarbons

c. Evaporation from fuel tank and carburetor accounts for about 15% of hydrocarbons.

As a result of vehicle modifications and changing legislation for automobile emissions through the 1960's, the magnitudes and relative contributions of these three sources vary with vehicle model year. Also, the manner in which each type of emissions is distributed in space must be treated individually. Exhaust and blow-by rates are computed on a grams per mile basis and may be distributed according to the spatial distribution of M_l over the urban area. To distribute evaporative losses in the same way, however, it is necessary to assume a daily mileage traveled by the average vehicle and a temporal distribution of evaporative losses. An alternative is to compute the total evaporative losses from all vehicles in the area and distribute these emissions in proportion to the non-freeway vehicle mileage driven in each gride square.

The emission rate of species k from the exhaust of an average vehicle in grams/mile is given by

$$Q_k = \sum_{i=1}^{m} x_i \sum_{j=1}^{N} Y_{ij}\, e_{ij}\, K_{ijk} \tag{16}$$

where subscript k designates

> CO (in %) when $k = 1$
> Hydrocarbons (in ppm) when $k = 2$
> NO$_x$ (in ppm) when $k = 3$

and

> x_i = fraction of total cars in the airshed of model year i; $i = 1,2,\ldots,m$
> T = total number of cars in the airshed
> Y_{ij} = fraction of $x_i T$ manufactured by company j; $j = 1,2,\ldots,N$
> e_{ijk} = volumetric emissions of species k from cars manufactured by maker j of model year i
> K_{ijk} = constant multiplier to convert ppm (or %) to grams per mile, a function of the average inertial weight of the car and the transmission type (manual or automatic)

For freeways and surface streets M_l is calculated from

$$M_l = d_l \sum_{u=1}^{s} n_u t_u \tag{17}$$

where

> n_u = vehicles per day on a particular road segment (given as traffic counts at a point)
> t_u = miles of road segment to which n_u is assigned
> d_l = fraction of daily (24 hour) traffic count assignable to hourly period l
> s = number of road segments contained in the grid square

Blow-by losses have been virtually eliminated in recent model year cars by the compulsory fitting of positive crankcase ventilation (PCV) valves. To estimate blow-by rates from vehicles in a particular area, we must refer to the laws concerning PCV devices in that area. Estimates of blow-by rates from various vehicles are given in Reference 51. Evaporative emissions are difficult to measure and only gross estimates may be made (*see,* for example, Reference 52).

More recently, Roberts *et al.* (53) have proposed a somewhat more sophisticated automotive emissions model. In particular, they have extended the model described to include both hot running and cold start modes of operation. Also vehicles traveling on freeways are assumed to be hot running whereas a distribution of emissions over time is assumed for a cold start trip. The effect of variations in the rate of vehicle starts on the average emissions rate is also included. This effect is a prime contributor to increased average emissions rates from an individual vehicle during the morning rush hours. Also included are variations in emissions rates as a function of average speed, for vehicles traveling on freeways. This effect is particularly important as it has been estimated that emissions from an individual vehicle, in grams per mile increase about 100% when a vehicle traveling on a freeway is slowed from 60 mph to a 20 mph "rush hour crawl." (*See* Reference 53 for full description of their model.)

Aircraft Emissions. Aircraft emissions may be represented in various ways. For example, a model that describes each aircraft departure and arrival might be considered. Such a model would be appropriate in a study limited to the estimation of concentration levels in the immediate vicinity of an airport. However, an aircraft emissions model suitable for inclusion in a comprehensive urban airshed model need not be so detailed. The model we shall describe here was formulated on the idea that it would be a part of an airshed model having a spatial resolution of the order of two miles and a temporal resolution of the order of several minutes to one hour.

The aircraft emissions model consists of two major parts: ground operations and airborne operations. Emissions from these operations are treated as lumped volume sources, generated in the cell into which they are injected. Ground operations consist of three distinct modes:

1. Taxi mode: between runway and satellite upon landing; between satellite and end of runway, waiting to take-off; and idle at satellite

2. Landing mode: from touchdown on runway to turn-off from runway

3. Take-off mode: start of take-off to lift-off.
Airborne operations are comprised of two modes:

1. Approach mode: descent from maximum vertical extent of airshed (*e.g.*, the base of an elevated inversion) to ground.

2. Climb-out mode: lift-off to attainment of upper vertical extent of airshed.

It is convenient to use the following assumptions to treat aircraft operations in the emissions model (These have been used for Los Angeles by Roberts *et al.* (50):

a. For every aircraft arrival, there is one departure. Furthermore, arrival and departure rates are equal in each time period.

b. Aircraft follow straight line flight paths from top of airshed to touchdown and lift-off to top of airshed.

c. The angles of ascent and descent of all aircraft at a particular airport are assumed to be fixed and equal to those angles associated with the type of aircraft (medium range jet transport, business jet, etc.) having the highest fraction of total operations at the airport. (A more accurate calculation would include angles of ascent and descent computed as the weighted sum of the angles assignable to each type of aircraft, weighted in proportion to the mass emission of each class.)

d. Flight paths originate and terminate at the most frequently used runway at each airport.

e. The proportion of aircraft of a given type that arrive and depart from each airport does not vary throughout the day.

Regarding aircraft emissions, the following assumptions seem reasonable:

a. Pollutants are emitted at a uniform rate during each of the five operating modes; rates are specific to each class of aircraft. For airborne operations, the mass of contaminants injected into a cell is proportional to the length of the flight path in that cell.

b. Aircraft emissions can be treated as continuous releases, emitted at a uniform rate and averaged over an appropriate time period.

GROUND OPERATIONS MODEL. The amount of contaminant species k emitted during ground operations equals the summed products of average emission rates and residence times in each of the three modes, taxi, landing and take-off. The rate of emissions of species k in grams per minute into any ground cell ijl because of ground operations for the hourly period l is given by

$$Q^k_{ijll} = \frac{0.454 \, d_l \alpha_{ij}}{60} \sum_{u=1}^{7} n_u M_u \sum_{g=1}^{3} f^k_{gu} C_{gu} \qquad (18)$$

where g is an index denoting the three ground operations modes (taxi, landing, take-off), u is an index denoting aircraft type:

1 = long-range jet transport
2 = medium-range jet transport
3 = business jet
4 = turboprop transport
5 = piston engine transport

6 = piston engine utility
7 = turbine engine helicopter

and

d_l = fraction of total daily flights in hour l
α_{ij} = fraction of airport area in cell i,j
n_u = number of flights/day of aircraft of type u
M_u = average number of engines/aircraft of type u
f^k_{gu} = pounds of pollutant k emitted per 1000 pounds fuel consumed by aircraft of type u operating in mode g
C_{gu} = pounds of fuel consumed per engine of aircraft of type u operating in mode g

AIRBORNE OPERATIONS MODEL. The mass of species k emitted into cell ijm during approach is assumed to be proportional to the length of the flight path occupying that cell. The corresponding rate of emissions is given by:

$$Q^k_{ijml} = \frac{10^{-3}d_l P_{ijm}}{K} \sum_{u=1}^{7} n_u M_u f^k_u C_u \frac{t_u}{t'_u} \tag{19}$$

where

t_u = time spent in descent from inversion height to touchdown by aircraft of type u
t'_u = time spent in descent from 3000 feet above ground elevation to touchdown by aircraft of type u
f^k_u = pounds of pollutant k emitted by aircraft of type u per 1000 pounds fuel consumed during descent
C_u = pounds of fuel consumed per engine of aircraft of type u during descent from 3000 feet above ground elevation
P_{ijm} = fraction of the length of the flight path assignable to cell ijm

The mass of species k emitted into cell ijm during *climb-out* is also given by (19), where t_u and f^k_u now apply to an aircraft ascending from lift-off to inversion height, and t_u and C_u to an aircraft ascending to 3000 feet above ground elevation.

Since all concentration units in the airshed model are expressed as parts per million (ppm), the following conversion formula is used:

$$c^k_{ijm}\left[\frac{\text{ppm}}{\text{min}}\right] = \frac{10^6 \, v Q^k_{ijm}}{V_{ijm}W^k} \tag{20}$$

where

Q^k_{ijm} = emissions rate of species k (pounds per minute)
W^k = molecular weight of species k
V_{ijm} = volume of cell ijm, cubic feet
v = 379 cubic feet per pound mole, the molal volume of an ideal gas at one atmosphere at 60°F

Major aircraft emissions studies have been reported by Lemke *et al.* (54), George *et al.* (55), Hochheiser and Lozano (56), Lozano *et al.* (57), Northern Research (58), Bastress and Fletcher (59), George *et al.* (60) and Platt *et al.* (61).

Fixed Source Emissions. Emissions from fixed sources may be represented in several ways, depending upon the temporal and spatial scales of interest and the aims of the overall simulation. Consider, for example, strong point sources, such as power plants and refineries. If one wishes to estimate concentrations only in the immediate vicinity of the source (or sources), a detailed description of plume transport and dispersion would be necessary. Several models of plume and puff behavior have been developed; these models are discussed by Turner (62) and Roberts *et al.* (63). In contrast, if one wishes to estimate average pollutant concentrations on a spatial scale of several square miles, it may be reasonable to consider fixed source emissions as well-mixed in the cell into which they are injected, thereby disregarding the behavior of individual plumes. Intermediate between these extremes is the case in which a detailed understanding of plume dynamics is not essential; a simple analysis is sufficient to determine the cell (or cells) in which emitted pollutants are to be uniformly distributed. This intermediate level of analysis is often appropriate in modeling urban airsheds when an inversion is present as the horizontal spatial scale is such that pollutants emitted from large fixed sources are distributed uniformly vertically in one to three characteristic horizontal lengths. All that is necessary in this case is to apportion properly the emitted material among a few downwind cells.

The authors have used this intermediate approach to treat power plants in the Los Angeles Basin modeling study (50). Some of these plants are situated along the coastline, and their emissions are advected across the Basin under prevailing wind conditions. Typically emissions from these plants travel a horizontal distance of 2–5 miles before they are considered well-mixed in the vertical. Since an individual cell is 2 miles by 2 miles and horizontal dispersion of the plume under low winds extends about 2 miles after a 2–5 mile traverse, the assumption of approximately uniform distribution immediately downwind of the source is reasonable. A computational scheme for apportioning emissions among cells downwind of the source under these circumstances is described by Roberts *et al.* (50).

The Treatment of Meteorology

Atmospheric transport and dispersion processes are expressed in an airshed model in numerous way. Wind speed and direction enter through the component variables, $\hat{u}$, $\hat{v}$, and $\hat{w}$ (or more generally $\hat{u}(x,y,z,t)$). The depth through which mixing occurs, often defined by the height of the base of an elevated inversion, must be known in order to specify boundary conditions. This depth can vary considerably with x,y,z, and t, as in the Los Angeles Basin. Finally, the turbulent eddy diffusivities, $K_H(x,y,z,t)$ and $K_V(x,y,z,t)$, appear in the continuity equations and in the boundary conditions. We now present a survey of approaches presently being used to specify these meteorological variables, and we review promising approaches that may be adopted in the future.

The Winds. The atmospheric boundary layer is usually characterized by complex flow patterns. The drag exerted on moving air masses by the earth's surface, coupled with the effects of variations in terrain, are largely responsible for the phenomenological effects observed. Moreover, in urban areas where man has erected large bodies of buildings of assorted sizes and shapes, the complexity of local flows is further enhanced. The heat island effect of the cities—the absorption, retention, and delayed release of energy to the atmosphere—also participates in shaping flow patterns. Thus, while large scale movements of air masses are always difficult to describe, the effects of surface features and of solid–fluid energy transfers on near-surface flows simply enhances the difficulty associated with specifying wind fields in the mesoscale and microscale.

Basically, there are two general approaches to the specification of the winds; one is based on physical modeling, the other on the correlation, interpretation, and analysis of field measurements. Physical modeling approaches involve numerical simulation of the turbulent atmospheric boundary layer, frequently through numerical integration of the coupled momentum and energy equations. Methods based on field measurements involve interpolation of wind speed and direction, both in space and time, using either simple rules or subjective judgment. Judgmental considerations form the basis for the construction of maps displaying streamlines and isotachs (contours of constant velocity). In principle, the former approach is fundamental, relies on data only to specify initial conditions and boundary conditions, and, when successfully developed, is universally applicable. In contrast, the latter approach is empirical, is based solely on data collected, and is limited in applicability.

While modeling approaches are inherently more desirable, empirical methods are presently the only approaches used. As indicated above, simulation of the atmospheric boundary layer is quite complex, requires substantial amounts of computing, and cannot currently predict with requisite accuracy. Our knowledge of turbulent diffusion, of the effects of terrain on flow patterns, and of energy transfer processes is insufficient now to permit accurate predictions. Investigators have adopted the more reliable, but more limited, methods of interpolation and map construction to specify wind fields. Here, we discuss both approaches—numerical simulation because we believe that as progress is made in developing boundary layer models and solution techniques, these methods will be of increasing interest, empirical estimation procedures because they represent the most suitable, practically available means for specification of the winds. Thus, a review of progress in boundary layer simulation and a summary of interpolation techniques that are commonly used by investigators today are given below.

NUMERICAL SIMULATION OF THE ATMOSPHERIC BOUNDARY LAYER. Numerical simulation of the atmospheric boundary layer is concerned with the solution of the continuity equation

$$\sum_{i=1}^{3} \frac{\partial u_i}{\partial x_i} = 0 \tag{21}$$

and the three-dimensional, time-dependent equations of motion. (For simplicity we assume that temperature is constant. In general, the equation of conservation of energy is also required to describe boundary layer dynamics.)

$$\rho \left(\frac{\partial u_i}{\partial t} + \sum_{j=1}^{3} u_j \frac{\partial u_i}{\partial x_j} \right) = \mu \nabla^2 u_i - \frac{\partial p}{\partial x_i} + \rho g \delta_{i3} \tag{22}$$

$$i = 1,2,3$$

where ρ is the fluid density, μ the viscosity, p the pressure, and δ_{i3} the Kronecker data for which $\delta_{ij} = 1$ for $i = j$ and $\delta_{ij} = 0$ for $i \neq j$. (Coriolis forces have not been included in (22) to simplify further the presentation.) Since atmospheric flow is turbulent, we decompose each velocity component and the pressure into a mean component and a stochastic component. After appropriately substituting into (21) and (22), we average the resulting equation to obtain the time-averaged equations

$$\sum_{i=1}^{3} \frac{\partial \bar{u}_i}{\partial x_i} = 0 \tag{23}$$

and

$$\rho \frac{\partial \bar{u}_i}{\partial t} + \sum_{j=1}^{3} \frac{\partial}{\partial x_j} (\rho \bar{u}_i \bar{u}_j) + \sum_{j=1}^{3} \frac{\delta}{\delta x_j} (\overline{\rho u'_i u'_j}) = -\frac{\partial \bar{p}}{\partial x_i} + \rho g \delta_{i3} \tag{24}$$

The viscous terms have been neglected in (24), implying that the Reynolds number is large and that the viscous sublayer is shallow and of little interest. We use an overbar in contrast to the brackets used previously to indicate the mean velocities since the averaging procedure here generally is different from that used for the concentrations.

We see, as for the species continuity equations, that when we perform the averaging procedure we obtain new dependent variables, in this case the terms $\overline{\rho u'_i u'_j}$. These new terms are usually called the turbulent momentum fluxes or the Reynolds stresses, but, as before, we have more dependent variables than equations. Thus, some means to evaluate the turbulent momentum fluxes must be developed. Although no rigorous method of obtaining a closed set of equations is known, a number of semi-empirical approaches have been proposed which yield qualitative or semi-quantitative results for appropriately chosen classes of problems.

The key aspect, then, in numerical simulation of the atmospheric boundary layer is the evaluation of the turbulent momentum fluxes in the time-averaged equations of motion (24). Considering this, we review briefly some of the more promising techniques that have been used to determine these fluxes. Our objective is not to give a full review, but rather to introduce the types of approaches which in the future may permit the solution of (23) and (24) and thus the prediction of urban wind fields.

The most common approach to evaluating the turbulent momentum fluxes is to assume that an eddy viscosity K_M exists such that

$$\overline{\rho u'_i u'_j} = - \rho K_M \frac{\partial \overline{u}_i}{\partial x_j} \tag{25}$$

The value of K_M depends on the properties of the mean flow at a particular location and time. To account for the contribution of thermal stratification (buoyancy) to the production or suppression of turbulent energy, K_M is taken to be a function of the local value of the flux Richardson number, which expresses the ratio of the rate of generation of energy by buoyancy forces to the rate of generation of energy by the turbulent momentum fluxes. In this approach the influence of the past history of the turbulence on velocity field is not considered; the approach is termed a local theory.

The use of local theories, incorporating parameters such as the eddy viscosity K_M and eddy thermal conductivity K_E, has given reasonable descriptions of numerous important flow phenomena, notably large scale atmospheric circulations with small variations in topography and slowly varying surface temperatures. The main reason for this success is that the system dynamics are dominated primarily by inertial effects. In these circumstances it is not necessary that the model precisely describe the role of turbulent momentum and heat transport. By comparison, problems concerned with urban meso-meteorology will be much more sensitive to the assumed mode of the turbulent transport mechanism. The main features of interest for mesoscale calculations involve abrupt changes in surface conditions to which the flow must adjust, primarily by means of these transport processes. The accuracy with which the assumed mechanism models predict the real flow system is now more important. For this reason the local theories which have been commonly accepted for larger scale or essentially unchanging flows in atmospheric circulation are now being reevaluated with respect to their applicability to the smaller (meso-) scale problems of interest in the urban atmosphere. Two classes of approaches have been proposed for these types of problems.

Lilly (64), Deardorff (65, 66), and others have pioneered a numerically based method which represents a partial compromise between purely local theories and more sophisticated approaches which are based on modeling the turbulent fluid as a homogeneous material with non-Newtonian viscoelastic behavior. In this approach a grid-scale averaging operator is applied to the governing equations with the averaging typically being over the grid volume of the calculations, thereby filtering out the subgrid scale (SGS) motions. Explicit calculations can be done for the filtered variables after assumptions are made for the SGS Reynolds stresses and turbulent energy fluxes which arise from the averaging process. The larger scale eddies of the turbulent flow are thus included without direct approximation in the computation, and the empirical assumptions are limited to those concerning the smaller scale motions of the turbulent flow. For atmospheric flows where heat transfer is negligible, Smagorinsky et al. (67), Lilly (64), and Deardorff (65, 66) have proposed forms for the SGS Reynolds stresses which are also local. However, the approximation so introduced is less severe than that made for the purely local models since it involves only the smallest scales of the turbulent flow. Reasonably good results have been obtained for the flow systems which have been investigated by approach.

The second class of techniques is based on the description of spatial and temporal variations of turbulent intensity (or kinetic energy) by a transport equation. Essentially, one regards K_M and K_E as being governed by a transport equation which describes the convection, diffusion, creation, and destruction of the turbulence. This approach has been pioneered by Daly and Harlow (68), Harlow and Nakayama (69), and Bradshaw et al. (70).

In the most systematic application of this approach, Harlow and co-workers at Los Alamos have derived a transport equation for the full Reynolds stress tensor $\rho \overline{u'_i u'_j}$. They have coupled this equation with a scalar dissipation transport equation and have utilized with various semi-empirical approximations to evaluate the numerous unknown velocity, velocity–pressure, and velocity–temperature correlations which appear in the formulation. While this treatment is fairly vigorous, extensive computations are required to solve the governing equations. (For further details, *see* References *68, 69, 71,* and *72.*)

The work of Nee and Kovasznay (*73*) and, more recently, of Nee (*74*) proposes a much simpler approach for two-dimensional flows, one in which a single transport equation is written for the full shear viscosity (molecular plus turbulent). Much of the simplicity of this approach is gained by introducing postulated relationships for the production and decay of viscosity. Although certain limitations are clearly inherent in this approach, the results of various tests suggest that the model may be sufficiently detailed to account for the most relevant mechanisms of the turbulent momentum transport.

In summary, while most studies of atmospheric boundary layer flows have used local theories involving eddy transport coefficients, it is now recognized that turbulent transport coefficients are not strictly a local property of the mean motion but actually depend on the whole flow field and its time history. The importance of this realization in simulating mean properties of atmospheric flows depends on the particular situation. However, for mesoscale phenomena that display abrupt changes in boundary properties, as is often the case in an urban area, local models are not expected to be reliable.

INTERPOLATION OF MONITORING STATION DATA. In most urban areas surface wind speed and direction data are collected through a network of meteorological monitoring stations distributed widely over the region. The density of stations is usually sufficiently high to give an adequate qualitative description of the general flow over the area at any time but is never sufficient to permit the precise representation of the wind field. An interpolation scheme, either visual or automatic, is necessary to construct the velocity field near the ground from the wind station readings.

Visual techniques usually involve the preparation of maps, displaying streamlines and isotachs (contours of constant wind speed) for appropriate time intervals, usually 1 hour. As data are often collected by different agencies within a region (*e.g.,* the National Weather Service, the local air pollution control agency, etc.), wind speed and direction averages frequently differ as to the time interval over which they are computed (1 minute, 10 minutes, 1 hour), the beginning of the averaging period (the half hour or the hour for hourly averages), and the percentage of time that data are recorded (continuously over the hour, 5 minutes during each hour, etc.). Thus, raw data must be carefully organized before map preparation. We must also be careful to ensure consistency between maps on an hour-to-hour basis. Roth, *et al.* (*75*) have reported their experience in the use of visual techniques in preparing wind maps for the Los Angeles Area.

The map preparation is a tedious, time consuming process. It is more convenient to calculate the horizontal wind components at a point interpolating field wind data. Wayne *et al.* (22), Wendell (76), and McCracken *et al.* (36) have adopted formulas of the form

$$
u = \frac{\displaystyle\sum_{k=1}^{N} \frac{u_k}{r^2_k}}{\displaystyle\sum_{k=1}^{N} \frac{1}{r^2_k}} \ , \ v = \frac{\displaystyle\sum_{k=1}^{N} \frac{v_k}{r^2_k}}{\displaystyle\sum_{k=1}^{N} \frac{1}{r^2_k}}
$$

where u and v are the interpolated values, u_k and v_k, the horizontal components of the measured wind field, and r_k the distance between the point at which the wind vector is to be estimated and the kth wind station. It is generally required that all stations lie within some maximum distance of the point in question and that the number of field sites N be no less than three.

While the use of interpolation formulas is simple, rapid, and convenient, cases frequently occur in which they are either inapplicable or inappropriate. For example, the use of these techniques to estimate winds at or near the boundary of a region is often questionable, particularly when data outside the region are unavailable. Effects of terrain irregularities on local flow paterns cannot be accounted for. And it is difficult to spot spurious values and to check for consistency in space and time if formulas are the only way to analyze wind data. Thus, even if interpolation formulas are used, it may still be necessary to prepare a visual representation of data for preliminary scrutiny and analysis.

Virtually no measurements are made of winds aloft. Lack of these data make it necessary to formulate an approach to estimate the winds aloft from surface readings. The following guidelines are used to develop such an approach:

1. The constructed wind field must satisfy $\triangledown \cdot \hat{u} = 0$ at all points.

2. The winds aloft, up to the base of the inversion, do not vary greatly in speed from winds at the surface; wind directions do not vary significantly between the surface and the inversion base. (The latter assumption is undoubtedly violated, but in the absence of data it is the safest assumption.)

3. Flow is smoother and less tortuous aloft than at the surface.

4. Flow through the inversion base is negligible. (Regions of wind convergence are an exception since a continuously rising flow of air disrupts the inversion layer.) However, air from within the inversion layer mixes with air in the surface layer as the inversion base rises.

5. Lateral winds more or less follow the contour of the surface of the inversion base.

6. In all other respects the flow aloft resembles that at the surface. A highly simplified method for estimating elevated flow patterns, based on these guidelines, has been developed and implemented by Roth (75). However, such efforts are primitive and will remain so until measurements of winds aloft are made regularly.

The Elevated Inversion Over Cities. The depth of the layer through which vigorous mixing takes place is often difficult to specify. This is not the case, however, when an elevated inversion is present over a city. The height of the inversion base, H, and thus the depth through which mixing occurs is determined from vertical temperature soundings made on a routine basis, usually once or twice daily, over a convenient location such as the local airport. In airshed modeling, however, variations in mixing depth with time must be taken into account, and in certain areas spatial variations in mixing depth are also interesting. Over cities along the California coast, for example, mixing depth at a particular instant in time can vary greatly with distance inland.

Theoretical approaches to the prediction of $H(x,y,t)$ would involve the solution of the boundary layer equations for coupled energy and momentum transport or, more simply, the solution of the energy equations in conjunction with a constructed wind field. The application of such approaches to the prediction of inversion height has not yet been reported. Now, empirical models offer the only available means to estimate H. For those areas where it is necessary only to account for temporal variations in H, interpolation and extrapolation of measured mixing heights may be sufficient. When it is important to estimate H as a function of x,y, and t, a detailed knowledge of local meteorology is essential.

A study in estimating $H(x,y,t)$ is that carried out for Los Angeles by Edinger (77) (also Edinger and Helvey (78)). During the summer of 1957 Edinger and another pilot/meteorologist flew more than 100 flights over a 45 day period, measuring vertical temperature profiles at seven widely separated locations in the Los Angeles Basin. By analyzing the data he collected, Edinger was able to make the following generalizations:

"The marine layer over the Los Angeles coastal plain during the daylight hours (a) is shallow at the coast and deep inland, (b) increases in depth early in the day and then becomes progressively shallower during the afternoon, and (c) reaches its maximum depth first at the coastal stations and later at the inland stations."

Furthermore, variations in the depth of the mixing layer as a function of location and time can be explained by three atmospheric phenomena:

(a) convergence or divergence of the horizontal wind within the layer

(b) dilution of the marine layer from above by the mixing of air from within the elevated inversion layer with the marine air

(c) advection of deeper or shallower layers of marine air into the area.

By considering each of the three mechanisms separately and then combining them, Edinger was able to develop a semi-quantitative model of changes in the depth of the marine layer. This work has served as the basis for the construction of maps showing hourly variations in H as a function of x and y for two particular days of interest (*see* Roth *et al.* (75)).

Eddy Diffusion Coefficients. The eddy diffusivities, K_H (x,y,z,t) and K_V (x,y,z,t), which depend on the turbulent structure of the atmosphere, are two of the more elusive quantities that must be estimated. They are not established through direct measurement; they must be calculated from observed data. Most of the data that have been acquired to determine K_V (or K_H) have been limited to the surface layer (79); few data are available for conditions under which an elevated inversion was present. As a result, relatively little guidance is available in the literature that can be used to estimate these parameters.

Some useful qualitative observations can be made, however, regarding the turbulent diffusivity. It is a function of local velocity, shear field, and lapse rate; the functional relationship between K_H and K_V and these variables is largely unknown. Generally, K_V increases approximately linearly with z near the ground. In the presence of an elevated inversion, however, we expect K_V to decrease with increasing z in the upper portions of the surface layer owing to suppression of vertical buoyant fluctuations near the inversion base. Finally, values of K_V vary from about 50 m^2/minute under stable conditions to about 5000 m^2/minute under unstable conditions. K_H is generally neglected in urban airshed models, and here we will not consider it further.

Eschenroeder and Martinez (21) reviewed the literature pertaining to the turbulent structure of the atmosphere and, based on this effort, proposed a trapezoidal profile for K_V. As an example, they report the following formulations for K_V for a 180 meter inversion base: 30 square meters/min at the ground, increasing linearly to a height of 80 meters, at which height $K_V = 50(\overline{u} + 5)$ square meters/min ($\overline{u}$ is the horizontal wind speed). K_V is held constant from 80–135 meters elevation, whereupon it decreases linearly to 30 square meters/min at the inversion base. Note that this relationship expresses the variation of K_V with z through the assumed form of the distribution and with x, y, and t through variations in the horizontal wind with these quantities. However, the effects of shear field and vertical temperature structure are not considered in the formulation.

Other formulations have been proposed for K_V, including that of O'Brien (*80*), which is based on a variation with z similar to the trapezoidal form. Much work is required, however, of a theoretical and an experimental nature to include properly the effects of independent variables not thus far considered, and as a result to establish better the functionality expressing turbulent diffusivity and its variations.

Acknowledgment

The authors wish to express their gratitude to Thomas A. Hecht and Philip J. W. Roberts for their contributions in the preparation of the photochemistry and source emissions sections, respectively.

Literature Cited

1. Smith, M. E., "Chemical Reactions in the Lower and Upper Atmosphere," Interscience, New York, 1961.
2. Reiquam, H., *Atmospheric Environment* (1970) **4**, 233.
3. Reiquam, H., "Sulfur: Simulated Long-Range Transport in the Atmosphere," *Science* (1970) **170**, 318.
4. Reiquam, H., "Preliminary Trial of a Box Model in the Oslo Airshed," *2nd Clean Air Conf.*, Washington (Dec. 10, 1970).
5. Ludwig, F. L., Johnson, W. B., Moon, A. E., Mancuso, R. L., "A Practical Multipurpose Urban Diffusion Model for Carbon Monoxide," Stanford Research Institute, Menlo Park (1970).
6. Turner, D. B., "A Diffusion Model for an Urban Area," *J. Appl. Meteorol.* (1964) **3**, 83.
7. Clarke, J. F., "A Simple Diffusion Model for Calculating Point Source Concentrations from Multiple Sources," *J. Air Poll. Control Assoc.* (1964) **14**, 347.
8. Miller, M. E., Holzworth, G. C., "An Atmospheric Diffusion Model for Metropolitan Areas," *J. Air Poll. Control Assoc.* (1967) **17**, 46.
9. Koogler, J. B., Sholtes, R. S., Danis, A. L., Harding, C. I., "A Multivariable Model for Atmospheric Dispersion Predictions," *J. Air Poll. Control Assoc.* (1967) **17**, 211.
10. Hilst, G. R., "An Air Pollution Model of Connecticut," p. 251, IBM Scientific Computing Symposium, 1967.
11. Slade, D. H., "Modeling Air Pollution in the Washington, D. C. to Boston Megalopolis," *Science* (1967) **157**, 1304.
12. Bowne, N. E., "A Simulation Model for Air Pollution over Connecticut," *J. Air Poll. Control Assoc.* (1969) **19**, 570.
13. Roberts, J. J., Croke, E. J., Kennedy, A. S., Norco, J. E., Conley, L. A., "A Multiple Source Atmospheric Dispersion Model," Argonne National Laboratory Report (1970).
14. Lamb, R. G., "An Air Pollution Model of Los Angeles," M.S. Thesis, Department of Meteorology, University of California, Los Angeles (1968).
15. Seinfeld, J. H., in "Development of Air Quality Standards," A. Atkisson and R. Gaines, Eds., Merrill, Columbus (1970).

16. Neiburger, M., Edinger, J. G., Chin, H. C., "Meteorological Aspects of Air Pollution and Simulation Models of Diffusion, Transport and Reactions of Air Pollution," Project Clean Air, Vol. 4, University of California (Sept. 1, 1970).

17. Roth, P. M., Reynolds, S. D., Roberts, P. J. W., Seinfeld, J. H., "Development of a Simulation Model for Estimating Ground Level Concentrations of Photochemical Pollutants," Systems Applications, Beverly Hills (June 1971).

18. Eschenroeder, A. Q., "An Approach for Modeling the Effects of Carbon Monoxide on the Urban Freeway User," General Research Corporation, Santa Barbara (January 1970).

19. Friedlander, S. K., Seinfeld, J. H., "A Dynamic Model of Photochemical Smog," Environ. Science Technol. (1969) 3, 1175.

20. Lamb, R. G., Seinfeld, J. H., "Mathematical Modeling of Urban Air Pollution—General Theory," Environ. Science Technol., in press.

21. Eschenroeder, A. Q., Martinez, R. R., "Mathematical Modeling of Photochemical Smog," General Research Corporation, Santa Barbara (December 1969).

22. Wayne, L. G., Danchick, R., Weisburd, M., Kokin, A., Stein, A., "Modeling Photochemical Smog on a Computer for Decision-Making," J. Air Poll. Control Assoc. (1971) 21, 334.

23. Douglas, J., Jr., DuPoint, T., "Galerkin Methods for Parabolic Equations," SIAM J. Numerical Analysis (1970) 7 (4), 575.

24. Crowley, W. P., "Numerical Advection Experiments," Monthly Weather Rev. (1968) 96, 1.

25. Fromm, J. E., "Practical Investigation of Convective Difference Approximations of Reduced Dispersion," Phys. Fluids Supple. II (1969) II-3.

26. von Rosenberg, D. U., "Methods for the Numerical Solution of Partial Differential Equations," American Elsevier, New York, 1969.

27. Forsythe, G. E., Wasow, W. R., "Finite Difference Methods for Partial Differential Equations," Wiley, New York, 1960.

28. Ames, W. F., "Numerical Methods for Partial Differential Equations," Barnes and Noble, New York, 1969.

29. Richtmyer, R. D., Morton, K. W., "Difference Methods for Initial Value Problems," 2nd Ed., Wiley-Interscience, New York, 1967.

30. Peaceman, D. W., Rachford, Jr., H. H., "The Numerical Solution of Parabolic and Elliptic Differential Equations," J. Soc. Ind. Appl. Math. (1955) 3, 28.

31. Eschenroeder, A. Q., Martinez, J. R., "Further Development of the Photochemical Smog Model for the Los Angeles Basin," General Research Corporation, Santa Barbara (March 1971).

32. Harlow, F. H., "Methods in Computational Physics," Vol. 3, B. Alder, S. Fernbach, and M. Ratenberg, Eds., Academic, New York, 1964.

33. Sklarew, R. C., "A New Approach: The Grid Model of Urban Air Pollution," APCA Paper 70-79 (1970).

34. Ulbrich, E. A., "Adapredictive Air Pollution Control for the Los Angeles Basin," Socio-Econ. Plan. Sci. (1968) 1, 423–440.

35. Seinfeld, J. H., "Mathematical Models of Public Systems," Simulation Councils Proc. (1971) 1, 1.

36. MacCracken, M. C., Crawford, T. V., Peterson, K. R., Knox, J. B., "Development of a Multi-Box Air Pollution Model and Initial Verification for the San Francisco Bay Area," 52nd Ann. Meet. Amer. Meteorol. Soc., New Orleans (Jan. 10-13, 1972).

37. Leighton, P. A., "Photochemistry of Air Pollution," Academic, New York, 1961.

38. Altshuller, A. P., Bufalini, J. J., "Photochemical Aspects of Air Pollution: A Review," Photochem. Photobiol. (1965) 4, 97.

39. Altshuller, A. P., Bufalini, J. J., "Photochemical Aspects of Air Pollution: A Review," *Environ. Science Technol.* (1971) **5** (1), 39.
40. Johnston, H. S., Pitts, Jr., J. N., Lewis, J., Zafonte, L., Mottershead, T., "Atmospheric Chemistry and Physics," Project Clean Air, Vol. 4, University of California (Sept. 1, 1970).
41. Wayne, L. G., "The Chemistry of Urban Atmospheres," Technical Progress Report III, Los Angeles County Air Pollution Control District (December 1962).
42. Eschenroeder, A. Q., "Validation of Somplified Kinetics for Photochemical Smog," General Research Corporation, Santa Barbara (1969).
43. Behar, J. V., "Simulation Model of Air Pollution Photochemistry," Project Clean Air, Vol. 4, University of California (Sept. 1, 1970).
44. Westberg, K., Cohen, N., "The Chemical Kinetics of Photochemical Smog as Analyzed by Computer," Aerospace Corporation, El Segundo (1970).
45. Hecht, T. A., Seinfeld, J. H., "Development and Validation of Generalized Mechanism for Photochemical Smog," *Environ. Science Technol.* (1972) **6**, 47.
46. Eschenroeder, A. Q., Martinez, J. R., "Analysis of Los Angeles Atmospheric Reaction Data from 1968 and 1969," General Research Corporation, Santa Barbara (July 1970).
47. Altshuller, A. P., "An Evaluation of Techniques for the Determination of the Photochemical Reactivity of Organic Emissions," *J. Air Poll. Control Assoc.* (1966) **16** (5), 257.
48. Bonamassa, F., Wong-Woo, H., "Composition and Reactivity of Exhaust Hydrocarbons from 1966 California Cars," 152nd National Meeting of the American Chemical Society, New York (Sept. 11-16, 1966).
49. Seinfeld, J. H., Hecht, T. A., Roth, P. M., "Development of a Simulation Model for Estimating Ground Level Concentrations of Photochemical Pollutants," Appendix B, Systems Applications, Beverly Hills (May 1971).
50. Roberts, P. J. W., Roth, P. M., Nelson, C. L., "Development of a Simulation Model for Estimating Ground Level Concentrations of Photochemical Pollutants," Appendix A, Systems Applications, Beverly Hills (March 1971).
51. Sigworth, H. W., Bureau of Air Pollution Sciences, Environmental Protection Agency, private communication, May 1971. (Emissions rates reported in [50].)
52. Hurn, R. W., "Air Pollution, Vol. III: Sources of Air Pollution and Their Control," 2nd ed., A. C. Stern, Ed., Academic, New York, 1968.
53. Roberts, P. J. W., Liu, M. K., Roth, P. M., "A Vehicle Emissions Model for the Los Angeles Basin—Extensions and Modifications," Systems Applications, Beverly Hills (April 1972).
54. Lemke, E. E., Shaffer, N. R., Verssen, J. A., "Air Pollution from Aircraft in Los Angeles County, Los Angeles County Air Pollution Control District (December 1965).
55. George, R. E., Verssen, J. A., Chass, R. L., "Jet Aircraft—A Growing Pollution Source," *Ann. Meet. Air Poll. Control Assoc.*, Paper 69-191 (June 1969).
56. Hochheiser, A., Lozano, E. R., "Air Pollution Emissions from Jet Aircraft Operating in the New York Metropolitan Area," *Air Trans. Meet. Soc. Automotive Engineers*, New York (April 29–May 2, 1968).
57. Lozano, E. R., Melvin, Jr., W. W., Hochheiser, S., "Air Pollution Emissions from Jet Engines," *J. Air Poll. Control Assoc.* (June 1968) **18**, 392.
58. Northern Research and Engineering Corporation, "Nature and Control of Aircraft Engine Exhaust Emissions" (November 1968).
59. Bastress, E. K., Fletcher, R. S., "Aircraft Engine Exhaust Admissions," *Amer. Soc. Mech. Engrs Paper*, Los Angeles (Nov. 16-20, 1969).

60. George, R. E., Nevitt, J. S., et al., "Study of Jet Aircraft Emissions and Air Quality in the Vicinity of the Los Angeles International Airport, Los Angeles County Air Pollution Control District (April 1971).
61. Platt, M., Bakir, R. C., Bastress, E. K., Chang, K. M., Siegel, R. D., "The Potential Impact of Aircraft Emissions upon Air Quality," Northern Research and Engineering Corporation Report 1167-1 (December 1971).
62. Turner, D. B., "Workbook of Atmospheric Dispersion Estimates," Public Health Service, Cincinnati (1970).
63. Roberts, J. J., Croke, F. J., Kennedy, A. S., Nores, J. E., Conley, L. A., "A Multiple Source Urban Atmospheric Dispersion Model," Argonne National Laboratory, Argonne (1970).
64. Lilly, D. K., "The Representation of Small Scale Turbulence in Numerical Simulation Experiments," p. 195, IBM Report 320-1951 (1967).
65. Deardorff, J. W., "A Three-Dimensional Numerical Study of Turbulent Channel Flow at Large Reynolds Numbers," J. Fluid Mech. (1970) 41, 453.
66. Deardorff, J. W., "A Three-Dimensional Numerical Investigation of the Idealized Planetary Boundary Layer," Geophys. Fluid Dynam. (1970) 1 (377), 377.
67. Smagorinsky, J., Manabe, S., Holloway, J., "Numerical Results from a Nine Level General Circulation Model of the Atmosphere," Monthly Weather Rev. (1965) 93, 727.
68. Daly, B. J., Harlow, F. H., "Transport Equations in Turbulence," Phys. Fluids (1970) 13, 2634.
69. Harlow, F. H., Nakayama, R. I., "Turbulence Transport Equations," Phys. Fluids (1967) 10, 2323.
70. Bradshaw, P., Ferriss, D. H., Atwell, N. P., "Calculation of Boundary-Layer Development Using the Turbulent Energy Equation," J. Fluid Mech. (1967) 28, 593.
71. Harlow, F. H., "Transport of Anisotropic or Low-Intensity Turbulence," Los Alamos Report LA-3947 (1968).
72. Harlow, F. H., Hirt, C. W., "Generalized Transport Theory of Anisotropic Turbulence," Los Alamos Report LA-4086 (1968).
73. Nee, V. W., Kovasznay, L. S. G., "Simple Phenomenological Theory of Turbulent Shear Flows," Phys. Fluids (1969) 12, 472.
74. Nee, V. W., "Mass Diffusion from a Line Source in Neutral and Stratified Surface Layers," presented at 1971 Annual Meeting, Division of Fluid Mechanics, American Physical Society, San Diego (1971).
75. Roth, P. M., Reynolds, S. D., Roberts, P. J. W., "Development of a Simulation Model for Estimating Ground Level Concentrations of Photochemical Pollutants," Appendix C, Systems Applications, Beverly Hills (June 1971).
76. Wendell, L. L., "A Preliminary Examination of Mesoscale Wind Fields and Transport Determined from a Network of Wind Towers," NOAA Technical Memorandum ERLTM-ARL 25 (1970).
77. Edinger, J. G., "Changes in the Depth of the Marine Layer over the Los Angeles Basin," J. Meteorol. (1959) 16, 219.
78. Edinger, J. G., Helvey, R. A., "The San Fernando Convergence Zone," Bull. Amer. Meteorol. Soc. (1961) 42, 626.
79. Lumley, J., Panofsky, H. A., "The Structure of Atmospheric Turbulence," Wiley, New York, 1968.
80. O'Brien, J. J., "A Note on the Vertical Structure of the Eddy Exchange Coefficient in the Planetary Boundary Layer," J. Atmospheric Sci. (1970) 27, 1213.

RECEIVED July 19, 1971.

4

Concepts and Applications of Photochemical Smog Models

ALAN Q. ESCHENROEDER and JOSE R. MARTINEZ

General Research Corp., Santa Barbara, Calif. 93105

Following an overview of mathematical methods of analyzing air pollution, detailed developments of inputs, techniques, and validations are presented for photochemical smog modeling. Finite-difference formulations are used to compute concentration histories. The chemical kinetics are expressed as lumped-parameter reaction mechanisms derived from published laboratory data. Turbulent diffusion coefficients, which depend on height and time, come from atmospheric measurements. Inputs consist of source inventories for the Los Angeles basin and solar irradiation curves for the appropriate days. Predicted histories of reactive hydrocarbons, oxides of nitrogen, and ozone are consistent with the variations observed at air monitoring stations. With refined descriptions of advection, the mathematical model will serve as a tool in planning legislation and guiding urban planning in the future.

Atmospheric simulation models will play a central role in our efforts to improve the quality of the urban air environment because decision-makers must anticipate the outcomes of available courses of action. Although air pollution models are still being developed, their authentication will provide the link needed to establish the source emission limits required to achieve desired air quality standards. This approach can be broadened to answer questions of social and economic significance for different ameliorative strategies.

The expected evolution of modeling methods leads ever closer to a direct contact with our immediate problems. We are presently seeking a technical understanding of chemical interactions with transport processes, and our next step is to integrate this understanding in cost-benefit analyses. (Because of the difficulties arising, probably a more realistic goal is a cost-of-restoration study.) Successive improvement and check out of modeling techniques will offer quantitative criteria for federal and state legislation to improve air quality by controlling mobile sources of emissions. Addressing more specific problems, local government needs criteria for urban planning to prescribe population patterns and industrial land uses that tend to preserve air quality. Hence, the local application of predictive modeling influences the dispositions of fixed and mobile emitters. At the finest level of detail a predictive model could give the logic for the data processor in a real-time tactical system to guide decisions for temporary source shutdowns that would avoid calling alerts after dangerous pollution buildups.

Although we hope that preventive policies would preclude the need for this kind of alarm system, it is possible that some communities may be moving toward unacceptable limits faster than controls can be imposed.

Implementation plans for abating air pollution problems require a quantitative cause-and-effect relationship to estimate future air quality in terms of source controls proposed for the region. Mathematical models, which are most likely to give this needed connection, must ultimately describe chemical transformations as well as meteorological transport. Air quality criteria now exist for particulate matter, sulfur oxides, carbon monoxide, photochemical oxidants, nitrogen oxides, and hydrocarbons. Each of these materials is involved in chemical changes in urban atmospheres. All, except possibly CO (carbon monoxide) and some particulates, are either transformed or decay within a short enough time so that their concentrations in the urban domain are substantially affected by chemical reactions. Although our understanding of the detailed mechanisms is far from complete, we now possess a workable mathematical framework for modeling atmospheric dispersion of multicomponent pollutant systems undergoing coupled chemical reactions.

Having recognized these modeling requirements imposed by implementation planning, how can we best exploit our available methods to meet the immediate needs? For a limited number of regions, we can test some of the simplistic linear relationships connecting pollution with emissions (1, 2). For those areas that have had many sensors in the field measuring air contaminants over the years, the relationships are assumed to be derived from plots of pollution levels vs. emission tonnages. Global background concentration has been suggested as an initial intercept point (on the pollution level axis) for the straight-line formulas.

For most regions, however, not even a simple quantitative rationale can be used to implement planning because neither time nor money is available to deploy a network of sensors and to collect a significant sample of monitoring data. This dilemma suggests another approach: the requirements that have been imposed can be met by instituting a deliberate effort to adapt second-generation mathematical models to the implementation-planning problem. Admittedly, unresolved conflicts in present experimental results may seem to indicate that this approach is premature, but the urgency of the situation likely justifies parallel activities. The adaptation of the mathematical framework can simultaneously proceed with laboratory investigations and field programs. As fundamental knowledge expands, the blanks in the framework can be filled. Meanwhile, the preliminary framework will already be examining atmospheric pollution relationships. Also, early attempts to solve planning problems with models will highlight specific deficiencies and will give valuable feedback to the research community.

The two-way interaction between modeling and data-gathering can be considered as mutual support between research activities and enforcement activities. The efforts put forth in abatement are thereby more directly related to research efforts put forth on meteorology, chemistry, and physics. With the emerging complexities of economic factors, the decision-maker must use techniques more subtle than massive rollback. The legislative approach based on flat prohibitions can be pressed to some extent, but it ultimately meets some stout barriers in the real world. A powerful combination of quantitative justification and technological innovation will be needed to penetrate these barriers. Optimization will be the key word; concern for computer expenditures will decrease as the dollar costs of damage caused by air pollution and of resources needed to control it are generally realized.

Air Pollution Models

Dispersion Models Based on Inert Pollutants. Atmospheric spreading of inert gaseous contaminant that is not absorbed at the ground has been described by the various Gaussian plume formulas. Many of the equations for concentration estimates originated with the work of Sutton (3). Subsequent applications of the formulas for point and line sources state the Gaussian plume as an assumption, but it has been rigorously shown to be an approximate solution to the transport equation with a constant diffusion coefficient and with certain boundary conditions (4). These restrictive conditions occur only for certain special situations in the atmosphere; thus, these approximate solutions must be applied carefully.

Before extensive application of these plume models, Frenkiel used a puff model to study photochemical smog in Los Angeles (5). A point source was assumed to be centered within each four-mile-square in the grid pattern extending over the basin. Nashville, Tenn. was the subject of some later papers by Pooler and Turner, who used Gaussian formulations (6, 7). Point sources were assumed in Reference 6, but crosswind line sources were assumed in Reference 7 to approximate the emission emanating from each area element. Checks against field measurements of SO_2 were made for each. Turner allowed for a first-order reaction in the gas phase to remove SO_2 from the air.

Extending the work to include NO_x (oxides of nitrogen), Clarke divided Cincinnati into 16 angular sectors, each having four radial zones (8). This grid was centered on a monitoring station. Reactions were allowed to deplete SO_2 but not NO_x. Koogler later added allowance for changing wind direction by periodically treating old plumes as new line sources (9). Also, the vertical confinement of an elevated inversion layer was approximated along with height variation of wind speed; evidently, the SO_2 was not allowed to react. Miller and Holzworth also considered a limited mixing layer by assuming a uniform vertical profile when the "effluent reaches the top of the mixing layer" (10). The method was applied to Los Angeles, Washington, and Nashville for SO_2 and NO_x. A seasonal or monthly average model was reported by Martin and Tikvart to be applied to CO and SO_2 concentrations (11). Following Turner's equations (7), a program was developed for the IBM 1130 computer. Also among the early papers was that of Lucas (12) which gave preliminary approaches to the SO_2-removal mechanism.

Analysis and tabulations of data to be used in Gaussian plume formulas are also available. The report for the St. Louis Dispersion Study (13) gives further insight into tracer-spreading over urban areas in contradistinction to open areas where many measurements have been made. Detailed working charts and numerous examples in Turner's workbook (14) aid practical estimation of atmospheric dispersions under the conditions outlined above.

For approximate calculations of either peak or average pollutant levels, a simple slab model can be used for the air overlying the urban area. The ground is the lower boundary, and the uppermost extent of mixing—e.g., an elevated inversion base—is the upper boundary of the slab. Uniform vertical profiles are assumed, and horizontal flow carries the polluted air away. This concept was suggested by Smith for order-of-magnitude estimates (15); it was also discussed in Wanta's review article (16) along with most of the Gaussian plume models referred to above.

Trying to overcome the limitations of the Gaussian plume assumptions, Lamb presented a Green's function approach to the solution of the transport equation with both area sources at the ground and volumetric sources caused by reactions (17). Two other features included in the model were time dependence and absorption at the ground. (The latter generalization of boundary conditions might be especially important for pollutants such as NO_x, SO_2, and O_3, which combine with surfaces or with moisture on surfaces.)

Recent work at Stanford Research Institute has been directed toward a simple and practical urban diffusion model for carbon monoxide (*18*). Its objectives include not only hour-to-hour predictions but also long-term climatological effects. Traffic distributions in time and space are inputs to the model, and concentration history at some receptor point is an output. This approach is still being developed and will be improved with tests against field data to account for peculiarities of sampling sites and their microclimates. Combining the philosophy of this approach with our methods of treating photochemical reactions may provide an urban modeling technique that could be used for planning and abatement.

Puff models such as that in Reference 5 use Gaussian spread parameters, but by subdividing the effluent into discrete contributions, they avoid the restrictions of steady-state assumptions that limit the plume models just described. A recently documented application of a puff model for urban diffusion was described by Roberts *et al.* (*19*). It is capable of accounting for transient conditions in wind, stability, and mixing height. Continuous emissions are approximated by a series of instantaneous releases to form the puffs. The model, which is able to describe multiple area sources, has been checked out for Chicago by comparison with over 10,000 hourly averages of sulfur dioxide concentration.

Related Work on Photochemical Smog Modeling. Models for photochemical air pollution require extensions of earlier methods. Coupled chemical reactions and radiation attenuation in the ultraviolet introduce nonlinearities into the analysis. Consequently, the superposition of linear solutions from collections of point, line, or finite-area sources may inaccurately describe the chemical interactions with meteorological conditions in the air basin. Chemical evolution of pollutants, therefore, demands a step-by-step description to reflect the cumulative effects of the processes occurring.

Efforts aimed at overcoming the limitations have resulted in methods that range from modified classical treatments to new formulations. Because detailed chemical descriptions were not yet available, the work of Frenkiel (*5*) on the Los Angeles basin stresses diffusive aspects of the problem by treating plume trajectories from a collection of puff sources. Other early work (*20*) used simplified chemistry in an analog computer solution for composition histories in a network of homogeneously mixed cells. The horizontal faces of each cell were assumed to be the inversion base and the ground, respectively. This approach somewhat resembles that in Reference *15*. Suggestions have been made that chemical changes should be superimposed on a dispersion model like Frenkiel's (*5*).

Friedlander and Seinfeld, recognizing that the generality needed could be given by finite-difference numerical solutions to the diffusion equations, developed their dynamic model of photochemical smog (*21*). This model allowed for nonlinear chemical interactions, but the Lagrangian profile similarity required by the diffusion scheme still may be too restrictive to allow full freedom in specifying ground boundary conditions. Tests of the technique for photochemical smog with actual emission inventory data (in the form of distributed ground boundary conditions) have not been reported.

The models developed by Calvert (*22*) and Wayne *et al.* (*23*) stress chemical kinetic mechanism. In the former the photochemical smog mechanism is reduced to 17 steps and in the latter to 31 steps. In the larger mechanism plausible reactions for the $C_3H_6/NO_x/$ air system are carefully described, but it is ultimately applied to the atmospheric case where dozens of smog-forming hydrocarbons are involved. To preserve a consistent level of detail, the number of reactions must be increased by a large factor; however, this is avoided by adjusting the rate constants to obtain observed results. Homogeneously stirred air parcels following wind trajectories are assumed to retain first order, ordinary differential equations. Chemical kinetic studies have been carried out by numerical integrations of the coupled rate equations for conditions appropriate to laboratory chamber experiments in the absence of diffusion. Other computer investigations of this general nature include the work of the group headed by Bernard Weinstock of Ford Motor Co. and Karl Westberg at the Aecrospace Corp. (*24*). These approaches follow the philosophy of studying a proposed chemical scheme by modeling laboratory data. Modifications are usually necessary for applications to the atmosphere.

Three-dimensional, time-dependent methods (*25, 26*) have been recently proposed, but results for reactive atmospheres have not been reported at this writing. Simplified chemistry must be used in each of these approaches because of the emphasis on details of advection and diffusion. The body of data for most air basins falls short of the input requirements for any transport formulation of this complexity. In some cases it may be difficult to avoid the problem of allowing too many unspecified parameters to obscure the physically based portions of the calculation. One new method uses an Eulerian coordinate frame (Lagrangian coordinates refer to a fluid mass which is followed in time and space in contrast with an Eulerian frame which has fluid moving relative to a fixed coordinate system.) which introduces the chronic problem of artificial (or numerical) diffusion. Small time and space steps are the only means suggested previously to remedy the problem. The other method uses a discretized representation of species concentrations in a Lagrangian frame. For multicomponent systems the latter approach has previously been limited because of the large quantity of rapid access memory storage needed for each species. Both of these problems have been recognized for some time with three-dimensional, time-dependent calculations. More powerful computers and novel algorithms may be required for their solutions.

Overview of the Combined Photochemical–Diffusion Model. Trying to incorporate the best aspects of some previous work, we have tried to balance the detail between the chemical and meteorological formulations in our model. The development methodology begins with pure chemical kinetics validations using laboratory data from irradiation chamber experiments. This development has evolved through two stages—one involving only seven reactions and another using 12 reactions. Both are based on a lumped-parameter approach that includes groups of parallel reactions that perform a common function into a single effective kinetic step. Another aspect of the lumping is the collapse of a reaction chain into a rate-controlling step. Nonstoichiometric product yields sometimes occur with these simplifications.

The application of the chemical schemes to atmospheric phenomena requires a diffusion formulation that reflects time-dependence and spatial variability of meteorological conditions. An attempt has been made to keep the mathematical description near the level of detail and precision of the observational data. This has resulted in a Lagrangian air parcel formulation with finite-rate vertical diffusion. The approach avoids the artificial numerical diffusion because it uses natural (or intrinsic) coordinates that are aligned with fluid motion. This allows us simultaneously to include upward dispersion and chemical change. Figure 1 schematically illustrates the main features of the formulation. High-speed memory requirements are limited by allowing sequential point-by-point output of the history of the air parcel.

Some insights into model requirements are gained by examining some length and time scales characteristic of the Los Angeles basin, the region chosen as a prototype for validation. The horizontal scale of the air region is tens of kilometers, but the distance from ground to inversion base is usually less than one-half kilometer during smog episodes. During late morning buildup of secondary pollutants, winds are only 3 or 4 km/hour, and the vertical diffusion coefficient averages about 10^{-2} km^2/hour or less. High reactivity hydrocarbons are halfway photo-oxidized about 2 or 3 hours after they appear in the morning. If vertical

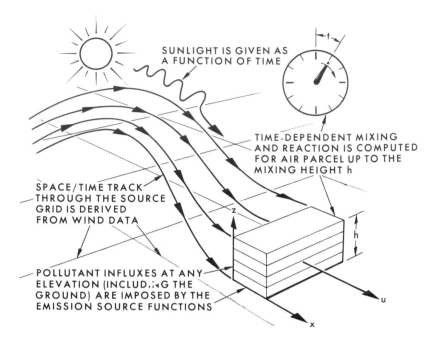

Figure 1.   Schematic of diffusion model for air pollution simulation

diffusion approximates a random walk, the characteristic diffusion time from the ground to the inversion base can be 1–4 hours. A 10 km downwind drift also occurs within this interval. Consequently, we conclude that a smog event must be modeled by treating simultaneously the processes of vertical diffusion, horizontal advection, and finite-rate photochemistry.

Some Simplified Kinetic Mechanisms

The Original Scheme. The photochemical reactions lead to most of the unusual new requirements placed on this model. Before treating atmospheric studies, we must understand the action of the chemistry and describe it simply enough to keep the ultimate computer requirements within reasonable bounds. First, an extremely simple version of a kinetic mechanism (27) will be explained and examined.

Experimental chamber results have been computer analyzed to give a kinetic model for the Los Angeles basin studies. In developing the model scheme, we began by trying combinations of elementary reaction steps that have been previously suggested (28, 29, 30, 31, 32, 33). After many successive simplifications, we obtained a seven species by seven reaction model mechanism which is the first version discussed. Simultaneous solutions of the coupled rate equations guided the evolution of the representation. We retained only those classes of rate-controlling steps needed to replicate observed concentration histories. However, enough generality is included to validate the assumptions over a range of composition and light intensity conditions representative of photochemical air pollution. Hence, no claim is made that the scheme is a complete, mechanistic embodiment of the available chemical kinetic hypotheses.

Figure 2 illustrates the lumped-parameter kinetics by a block diagram, and Figures 3 and 4 show concentration histories in a *trans*-2-butene/nitric oxide system which illustrate the action. Symbolically, the diagram shows species as boxes and reactions as path intersections. In Figure 2 the emitter is labeled SOURCE and sunlight is denoted by $h\nu$. The experiment (*34*) illustrated in Figures 3 and 4 replaces the source with a premixed charge of HC and NO and uses simulated sunlight.

Examination of Figure 2 shows that the NO_2 photodissociates to contribute O atoms and NO. The O atoms attack the hydrocarbon and

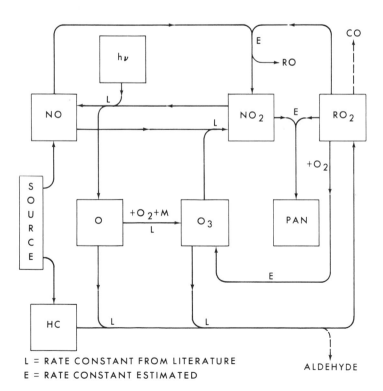

L = RATE CONSTANT FROM LITERATURE
E = RATE CONSTANT ESTIMATED

Figure 2.   Original reaction mechanism

produce free radicals represented here by RO_2. In turn, the RO_2 provides an oxygen source to reconvert NO to NO_2. This path augments the already rapid reaction of ozone with NO to form NO_2. Three-body association of O with O_2 contributes most of the ozone; however, an abstraction reaction of O_2 with RO_2 also gives ozone. (In the more advanced version of the model this abstraction reaction is dropped because it is energetically unfavorable. It is needed here to provide ozone buildup at late times.) Nitrate formation removes NO_2 in chain termination reactions. One minor branch of the chain is included in Figure 2 as the production of peroxyacetyl nitrate (PAN). This is included in the model because PAN has been established to be a phytotoxicant.

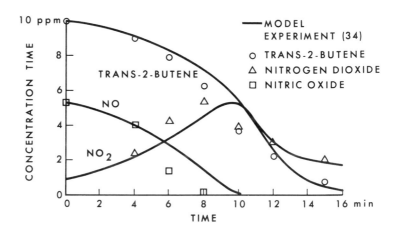

Figure 3.   Computed and observed reactant histories for the trans-*2-butene/nitric oxide system*

Following the concentration histories in Figures 3 and 4, we note that the hydrocarbon disappearance contributes RO_2 which oxidizes NO to NO_2. When this induction phase is complete, ozone concentration rises because no NO remains to compete for the ozone. Nitrate formation and photodissociation then take over to reverse the rising trend in NO_2. In the later stages ozone becomes a serious competitor with oxygen atoms in the attack on hydrocarbons.

Before we assess the validity of the calculations, we shall list some of the processes which are absorbed in certain overall steps so that the simplicity of the scheme is preserved for ultimate economy in computation:

1. Free radicals like RO_2 or OH also attack hydrocarbons so that the O + HC and O_3 + HC rates must be artificially elevated.

2. Aldehyde photolysis (35) is another branch of the mechanism that must be absorbed in the rates and stoichiometry of a lumped parameter scheme.

3. Inorganic particulate nitrate products would be added if it were necessary to complete the material balance to account for nitrogen (36).

4. Radical regeneration involving CO may be implicitly absorbed in the chain branching factor applied to RO_2 (37).

Curves in Figures 3 and 4 show calculated results (Quasi-stationarity is assumed for O-atom and RO_2-radicals to facilitate calculations. Numerical tests confirm the validity of these approximations.) for the conditions of one of Tuesday's experiments. Table I details the reaction stoichiometry and rate constants for the calculation. The calculated photooxidation rates for the main reactants are somewhat slow around 8 minutes, and the appearance rate of PAN is too fast between 12 and 16 minutes. This results from a compromise needed because of a certain inflexibility in the schematic kinetics. The calculation uses established values of rates from the literature for the three inorganic reactions. The NO_2 photolysis was calibrated (34) according to the light intensity at 0.37 min^{-1}, and the other two reaction rate constants were obtained from Reference 38, assuming an ambient temperature of 27°C. To obtain observed oxidation rates, the O atom plus C_4H_8 rate was increased nearly fourfold to absorb the competing radical reactions, and the ozone on C_4H_8 rate fell near the upper end of measured range. Nominal values for these reactions are taken from the literature (31). Collision theory estimates yield upper limit nominal values for the last three reactions listed in Table I.

A single experiment is discussed to show the validation of the simple scheme. Other experiments point out special features; for example, the

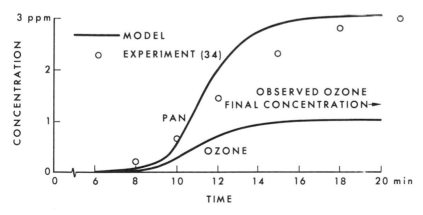

Figure 4. Computed and observed product histories for the trans-2-butene/nitric oxide system

Reaction	Table I. Rate Coefficients for *Nominal Values* [b]
$h\nu + NO_2 \rightarrow NO + O$	0.072 to 0.55 min^{-1}
$O + O_2 + M \rightarrow O_3 + M$	1.32×10^{-5} ppm^{-2}min^{-1}
$O_3 + NO \rightarrow NO_2 + O_2$	21.8 ppm^{-1}min^{-1}
$O + C_4H_8 \rightarrow 2.5\ RO_2 + 0.12\ Ald$	3.34×10^4 ppm^{-1}min^{-1}
$O_3 + C_4H_8 \rightarrow 2.5\ RO_2 + 0.12\ Ald$	4.2×10^{-2} to 6.4×10^{-1} ppm^{-1}min^{-1}
$RO_2 + NO \rightarrow NO_2 + RO$	≤ 50 ppm^{-1}min^{-1}
$RO_2 + O_2 \rightarrow 0.5\ O_3$	≤ 50 ppm^{-1}min^{-1}
$RO_2 + NO_2 \rightarrow 0.67\ PAN$	≤ 50 ppm^{-1}min^{-1}

[a] Stoichiometry imbalances occur in some reaction steps because of lumped parameter kinetic assumptions.

ozone reaction with a hydrocarbon is far more complex than a single step because somewhat lower rate values for $O_3 + C_4H_8$ (still in the nominal range shown in Table I) are observed in experiments involving C_4H_8/NO_2 systems under irradiation. Except for this complication, the rates in Table I show good modeling results near the value of initial C_4H_8/NO-ratio used in the prototype experiment. These rates also describe well the influence of varying light intensities below and above that used for Figure 3 (*see* Table I for range).

Several features of this particular system, however, are atypical of photochemical air pollution. Such features include the reactivity, the concentration levels, the time scales, and the organic nitrate production. *Trans*-2-butene is far more reactive than the average of atmospheric hydrocarbons. Even omitting low reactivity components (methane, ethane, and acetylene) from automotive exhaust compositions, we find that the remaining fraction of hydrocarbons has a weighted average reactivity index (39) many times less than that of the butenes. Regarding concentration levels, the experiment (34) cited begins with 10 parts per million (ppm) reactive hydrocarbon whereas atmospheric values are one tenth as large. As a result of these two differences, smog reaction times are about twenty times as long as those of the experiment. Finally the PAN yields from butene oxidation exceed those of the atmospheric reactants. Despite these limitations, Reference 34 covers a broader scope of parametric variables than most experiments reported in the literature. Therefore, it gives many validation benchmarks for various compositions and light intensities. Since the propylene–nitric oxide system is more nearly like reactive systems in air pollution, validation tests were conducted on that system also. Measured composition behavior from chamber irradiation tests (40, 41) again serves as a basis for modifying some of the rate constants to absorb the additional effects. Lower oxidation

Trans-2-Butene/Nitric Oxide System[a]

Model Values[c]

0.072 to 0.55 min^{-1}
1.32 $\times$ 10^{-5} ppm^{-2}min^{-1}
21.8 ppm^{-1}min^{-1}
1.11 $\times$ 10^{5} ppm^{-1}min^{-1}
6 $\times$ 10^{-1} ppm^{-1}min^{-1}

50 ppm^{-1}min^{-1}
3 $\times$ 10^{-5} ppm^{-1}min^{-1}
1 ppm^{-1}min^{-1}

[b] For sources *see* Some Simplified Kinetic Mechanisms.
[c] Values consistent with experimental measurement (*34*).

rates and lower PAN yields occur with propylene as compared with *trans*-2-butene. Lumped rate constants for O + C$_3$H$_6$ and O$_3$ + C$_3$H$_6$ were 4.97 $\times$ 10^4 and 1.8 $\times$ 10^{-1} ppm^{-1} min^{-1}, respectively, for use with this simple mechanism. In the more complex mechanism to follow the literature values are adhered to.

In viewing the transition from the simple to the complex mechanisms, one should realize what the limitations and objectives are for the simple mechanism. It describes pure hydrocarbon photooxidation with nitric oxide within a rather restricted range of initial mixtures typical of polluted air; however, it describes well the overall rate dependence on light intensity. The rate constants for organic reaction steps must be changed for different hydrocarbons as evidenced by the changes cited above needed to fit the propylene results. Major objectives of the more complex treatment discussed in the following sections are to relax the restrictions on initial mixture composition and to describe the chain initiating steps individually (instead of lumping them in the O + HC-rate).

Addition of More Realistic Oxidation Chains. As referred to above, it has been known for some time that hydrocarbon consumption rates observed in chamber experiments cannot be explained by ozone and oxygen atom attack alone. This is discussed by Altshuller and Bufalini (*31*) where they note that Schuck and Doyle (*42*) termed the disparity an excess rate. Our previous modeling work treated this by increasing the O$_3$ and O-atom rates of reaction with hydrocarbon.

To use known rate constants to a maximum extent in the present work, we have added a hydrocarbon oxidation chain to reflect the attack of RO upon the hydrocarbon. Because of its suspected dominance (*43, 44*), hydroxyl radical (OH) was assumed to be the only RO reacting with the hydrocarbon.

Following the inorganic cycle formed by

$$h\nu + NO_2 \rightarrow NO + O \quad \left.\begin{array}{c} \\ \\ \end{array}\right\} \tag{1}$$
$$O + O_2 + M \rightarrow O_3 + M$$

$$NO + O_3 \rightarrow NO_2 + O_2 \tag{2}$$

(The two reaction steps in (1) are combined by assuming O-atom stationarity internal to the logic of the computer program. Thus, Reaction (1) nets the system $h\nu + O_2 + NO_2 \rightarrow NO + O_3$.) We have allowed each hydrocarbon oxidation to generate multiple RO_2 radicals, expressed by the b-factors in the steps

$$O + HC \rightarrow b_1 (RO_2) \tag{3}$$

$$OH + HC \rightarrow b_2 (RO_2) \tag{4}$$

$$O_3 + HC \rightarrow b_3 (RO_2) \tag{5}$$

Subsequent conversion of NO to NO_2 was assumed to occur *via*

$$RO_2 + NO \rightarrow NO_2 + d (OH) \tag{6}$$

where d expresses that fraction of the conversion responsible for returning hydroxyl to the system.

Reactions (4) and (6) form something of a closed loop, and a stability requirement like $b_2 d < 1$ may be needed to prevent RO_2-runaway. This is not a precise requirement because of the variety of external factors that influnce the flow rate of free radicals around the loop. Some of the RO_2 is formed by the other two reactions, and there are radical removal processes in the chain termination steps we discuss below.

RO_2 and OH are held in check by removal steps that terminate the chains. We continue to use some lumped parameter reactions such as

$$RO_2 + NO_2 \rightarrow c (PAN) \tag{7}$$

and some elementary reactions like

$$OH + NO + M \rightarrow HONO + M \tag{8}$$

$$OH + NO_2 + M \rightarrow HNO_3 + M \tag{9}$$

This rounds out the expanded mechanism (*See* Figure 5 for a schematic diagram).

In summary, the main changes from the scheme described in The Original Scheme are the addition of radical species RO (treated as OH), the provision for a multiplicity of hydrocarbons (not shown in equations), and the elimination of ozone production from an RO_2 reaction. The multiplicity of hydrocarbons is done by adding parallel reactions to (3), (4), and (5). The ozone production reaction is omitted because of its suspected endothermicity (*24*).

For our initial tests we tried the rate constants compiled and estimated by Westberg and Cohen (*24*). Stoichiometric coefficients b_1, b_2, b_3, and d were determined by hand-calculation analysis of some of the propylene data represented by Reference *40*. We found that $bd \cong 1$ where b is a composite branching factor and $b \cong 2$ to explain the decay rate of NO as compared with that of propylene.

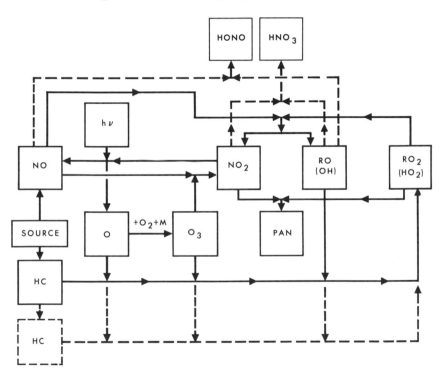

Figure 5.   Expanded kinetic mechanism

Explosions of RO_2-concentration characterized some early choices of bd-combinations because of the positive feedback loop involving radical reactions shown in Figure 5. Since these occurred in the first few seconds of real time, they were difficult to detect. This is unexpected because of the longer time scales usually associated with the system.

Table II.   Rate Coefficients for Expanded Model

Reaction	*Nominal Values for Propylene System* (24)
$h\nu + NO_2 \rightarrow NO + O$	0.4 min^{-1}
$O\ (+ O_2) + M \rightarrow O_3 + M$	1.32×10^{-5} ppm^{-2}min^{-1}
$O_2 + NO \rightarrow NO_2\ (+ O_2)$	22–44 ppm^{-1}min^{-1}
$O + HC \rightarrow 2RO_2$	6100 ppm^{-1}min^{-1}
$OH + HC \rightarrow 2RO_2$	244 ppm^{-1}min^{-1}
$RO_2 + NO \rightarrow NO_2 + 0.5\ OH$	122 ppm^{-1}min^{-1}
$RO_2 + NO_2 \rightarrow PAN$	122 ppm^{-1}min^{-1}
$OH + NO \rightarrow HNO_2$ [b]	99 ppm^{-1}min^{-1}
$OH + NO_2 \rightarrow HNO_3$ [b]	300 ppm^{-1}min^{-1}
$O_3 + HC \rightarrow RO_2$	$0.00927 - 0.0125$ ppm^{-1}min^{-1}
$(H_2O\ +)\ NO + NO_2 \rightarrow 2HNO_2$ [c]	
$h\nu + HNO_2 \rightarrow NO + OH$	

[a] Stoichiometry imbalances may occur because of lumped parameter assumptions.
[b] Rate constant lumps third body concentration.

Sudden drops of NO and HC (to some nonzero levels) occurred during the first minute followed by the dynamic equilibrium between NO, NO$_2$, and O$_3$ with a slow decay of hydrocarbon. The underlying pathology was discovered by rerunning with special diagnostic outputs.

Nominal rate constants were adjusted to reconcile computed results with measured values. For example, the (OH + HC)-rate was cut to about one-third the estimated value, and the (NO + RO$_2$)-rate was increased eightfold. Reaction rates which have been reported individually in the literature were held at their nominal values, including the (O + HC)-rate which we had to increase many-fold in The Original Scheme. Also, the retention of Reaction (5) was still necessary to describe the continued C$_3$H$_6$-decay after the near disappearance of NO.

To reproduce the late-time behavior of NO$_2$ and propylene, we included chain termination reactions in addition to those indicated in Figure 5. In their more recent review (45), Altshuller and Bufalini point out that HONO might be formed by the reaction with water vapor

$$NO + NO_2 + H_2O \rightarrow 2HONO \tag{10}$$

which is likely to proceed in the two steps

$$NO + NO_2 \rightarrow N_2O_3 \tag{11}$$

$$N_2O_3 + H_2O \rightarrow 2HONO \tag{12}$$

A possible source of OH-radical (43, 44) is the photodissociation of HONO

of the Hydrocarbon/Nitric Oxide Mechanism[a]

Model Values

0.4 min^{-1}
1.32 × 10^{-5} ppm^{-2}min^{-1}
40 ppm^{-1}min^{-1}
6100 ppm^{-1}min^{-1}
80 ppm^{-1}min^{-1}
1500 ppm^{-1}min^{-1}
6 ppm^{-1}min^{-1}
10 ppm^{-1}min^{-1}
30 ppm^{-1}min^{-1}
0.0125 ppm^{-1}min^{-1}
0.01 ppm^{-1}min^{-1}
0.001 min^{-1}

[c] Water vapor lumped into rate coefficient.

$$h\nu + HONO \rightarrow OH + NO \qquad (13)$$

Figure 6 shows the computed concentration history using this scheme to describe the experimental observations of Altshuller *et al.* (*40*), and Table II summarizes the reaction rate constants that were used. The Nominal Values column in Table II gives our original values, based on tabulations in Reference *24*. Following an empirical procedure, our calculations combined H$_2$O-concentrations with the rate constant for Reaction (10) as indicated; therefore, no comparable nominal value is shown. Higher levels of precision than that shown in Figure 6 were not among our objectives to describe the miscellaneous hydrocarbons in the urban atmosphere. Orders of magnitude for conversion times and levels are sufficiently good to proceed from here. The objective that we have fulfilled is to represent the so-called excess rate of hydrocarbon oxidation with added reactions instead of artificially increasing the O-atom rate with hydrocarbon.

Influence of Initial Composition

Influence of Initial Composition. SYNERGISM BETWEEN HYDROCAR-BONS. The fact that free radicals produced in the photooxidation of one hydrocarbon can accelerate the attack on a coexistent hydrocarbon has been illustrated with experimental findings by Altshuller and Bufalini (*45*). Because of the limited understanding of detailed interactive mechanisms, we limited our multiple hydrocarbon investigation to a simple parametric study. Using an abbreviated version of the mechanism in Table I, we added a parallel series of oxidation steps. This scheme is shown in the flow chart of Figure 5 with dashed lines along the bottom portion of the diagram.

In the parametric study the same initial values of total hydrocarbon and nitric oxide as those in Figure 6 were used. The hydrocarbon, however, was partitioned into two hypothetical compounds, one having triple the $(OH + HC)$ rate constant and the other having one-third the $(OH + HC)$ rate constant as propylene. The compound with the tripled rate is called species B, and the one with the decreased rate is called species A. All other reactions in the scheme are the same in all respects as their counterparts in the $(C_3H_6 + NO_x)$ system.

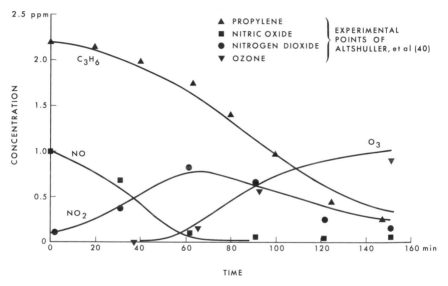

Figure 6.   Computed curves compared with experimental points for the propylene/nitrogen oxides system

Figure 7 shows the hydrocarbon decay curves from the parametric study. On the same coordinates we plot curves for model calculations of pure hydrocarbon A, pure hydrocarbon B, and pure propylene for comparison with the half-and-half mixture of hypothetical compounds. The most interesting aspect of Figure 7 is that the mixture decays faster than propylene even though it contains equal quantities of A and B which bracket propylene symmetrically. The coupling between the two hydrocarbons occurs *via* the RO_2 conversion of NO to NO_2 and the production of OH. Evidently the combination of the rapid incubation (afforded by the reactive hydrocarbon) with the twofold chain-branching is sufficient to increase the decay rate over that of pure propylene.

Figure 8 shows corresponding curves for ozone buildups predicted in the computer experiment. Again the combination seems to exceed the pure propylene in its rapidity to produce ozone. This occurs because of the enhanced conversion rate of nitric oxide to nitrogen dioxide stimulated by the early abundance of RO_2 generated by the high-reactivity fraction.

These findings show a plausible mechanism for the interaction of multiple hydrocarbons. In future applications of our photochemical/ diffusion model, this type of representation is probably more realistic than a super-detailed single hydrocarbon mechanism with many adjustable constants. Other radical interactions like RO_2 + HC and RCO_3 + HC might give plausible coupling paths (45) in addition to the hydroxyl attack on the hydrocarbon. Other forms of coupling may favor the slower reacting hydrocarbon in a similar numerical experiment with a half-and-half mixture.

MODELING THE INFLUENCE OF NO_x/HC-RATIO ON OXIDANT PRODUCTION. We included the chain termination results for nitrogen oxides above to improve the modeling of initial mixture effects. It is recalled that we previously issued a warning restricting the seven-step mechanism to a narrow range of NO_x/HC ratios. Seinfeld (46) pointed out that the slope of the curve of peak oxidant *vs.* NO_x/HC-ratio did not even have the correct sign. If modeling is to be used to evaluate hypothetical abatement strategies, this deficiency must clearly be remedied.

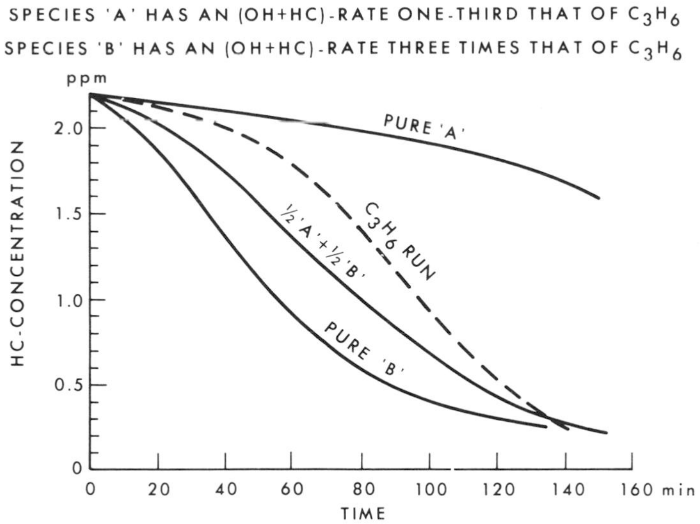

SPECIES 'A' HAS AN (OH+HC)-RATE ONE-THIRD THAT OF C_3H_6
SPECIES 'B' HAS AN (OH+HC)-RATE THREE TIMES THAT OF C_3H_6

Figure 7. Hydrocarbon decay with two reactivities

Figure 9 shows the results from the 12-step (Stationary state treatment of O-atoms reduces it to 11) mechanism given in Table II. Experimental results for pure and mixed hydrocarbons are shown on this plot of peak oxidant *vs.* NO_x/HC-ratio. The model mechanism shows a monotonically declining trend of oxidant production as the oxides of nitrogen increase. Both of the pure hydrocarbons measured by Altshuller *et al.* show peaks at widely separated values of the mixture ratio. It seem reasonable that the typical mixture of atmospheric pollutant hydrocarbons should show a decrease of oxidant with increasing NO_x/HC-ratio. This is supported by the fact that the relative abundance of hydrocarbon species decreases with increasing reactivity. Now since the low reactivity species peak at lower NO_x/HC-ratios the combined effect should give a negative slope.

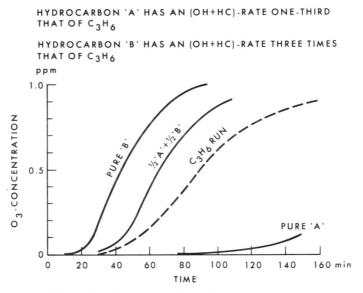

Figure 8.   *Ozone buildup with two reactivities*

The two plots on Figure 9 of oxidant production from diluted and irradiated automobile exhaust (48, 49) have negative slopes comparable with that of the model. As stated earlier, our purpose in using chamber data for propylene to validate the mechanism is to capture the main features of the experiments that apply to the atmosphere. However, if propylene were the subject of our study, we would proceed to expand on the mechanism to get the proper curve shape shown in Figure 9.

Having subjected the 12-step mechanism to these laboratory tests, we use it in its present form for atmospheric validation studies described in a later section.

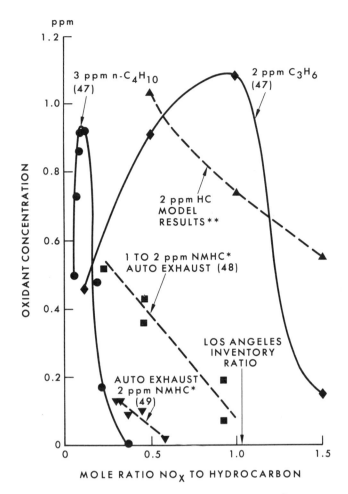

*Figure 9.   Influence of NO_x:HC-ratio on oxidant pro-
duction*

*This curve is lowered considerably by our subsequent adop-
tion of one-half the propylene rates for atmospheric calculations (NMHC = Non-methane hydrocarbon)*

EXAMINATION OF CARBON MONOXIDE EFFECTS. Recent work (*44,
50, 51*) of smog photochemistry has suggested that carbon monoxide
may play a role in accelerating the photooxidation of hydrocarbon/nitric
oxide mixtures. The mechanism suggested is

$$OH + CO \rightarrow CO_2 + H \tag{14}$$

$$H + O_2 + M \rightarrow HO_2 + M \tag{15}$$

$$HO_2 + NO \rightarrow OH + NO_2 \tag{16}$$

$$HO_2 + NO_2 \rightarrow HNO_2 + O_2 \tag{17}$$

Reaction (15) is postulated to contribute HO_2 rapidly as it responds to the presence of H-atom, and Reaction (16) enhances the NO-to-NO_2 conversion already occurring in Reaction (6). Likewise, the hydroxyl radical supplied by Reaction (16) goes on to react with the hydrocarbon in Reaction (4), producing RO_2-radicals in a branched chain already described. Reaction (17) is a chain termination step suggested for HO_2.

The key feature we wish to examine in this alternate path is the competition of CO with hydrocarbon for the hydroxyl radicals. The rate constant for $OH + C_3H_6$ suggested in Reference 24 is about equal to that suggested in Reference 50 for $OH + CO$. We tried a wide range of ($OH + CO$)-rates holding the basic system at the values in Table I. Since Reaction (15) is fast, Reactions (14) and (15) were added to give the overall reaction

$$O_2 + OH + CO \rightarrow CO_2 + HO_2 \tag{18}$$

The rate constant for Reaction (16) was taken to be the same as its RO_2-analog, Reaction (6), and the rate constant for Reaction (17) was assigned the value of 1.2 ppm^{-1} min^{-1}, based on the estimate in Reference 24. The species O_2 is implicitly carried in the calculation, and the rate constant for Reaction (18) is varied parametrically for this study.

Table III.   Influence of 100 ppm Carbon Monoxide on the System Shown in Figure 6 with Various Values of the Ratio of ($OH + HC$)/ ($OH + CO$) Reaction Rate Coefficients (Reaction Time in Minutes)

k_4/k_{18}	$t_{1/2}$-NO	t_{pk}-NO_2	$t_{1/2}$-HC	$t_{1/2}$-O_3
0.4 [a]	<1	4	40	10
4	6	13	48	22
40	25	43	73	62
80	29	52	81	73
System in Figure 6 (No CO)	36	66	92	83

[a] This is with k_{18} set at the estimated value from Reference 50.

Beginning with ($CO + OH$)-values at the level suggested in Reference 50 ($k_{18} \cong 200$ ppm^{-1} min^{-1}), we found rapid NO-to-NO_2 conversion and HC disappearance with 100 ppm CO added. Only when k_{18} was lowered to 1 or 2 ppm^{-1} min^{-1} did we obtain only small effects. Since much experimental verification is yet to be done, only a limited effort was given to this aspect of the study. Table III summarizes the main findings. It indicates that the half disappearance time of NO is the strongest effect of k_{18} on the system, and the half-disappearance time of hydrocarbon is least affected. The peaking time for NO_2 and the half-rise time of ozone are moderately sensitive to the rate constant.

In any event whether we use the estimated rate of (OH + CO)-reaction or only preserve its ratio to that of the (OH + HC)-reaction suggested in Reference 24, we get large effects. Other work (43) suggests a much larger (OH + HC)-rate constant for propylene. Because of our findings and of the wide disagreements on some of the rates, a real need exists for definitive experimentation to achieve a deeper understanding of the kinetics.

Atmospheric Adaptation Based on Reactivity Data Analysis. To apply the reaction mechanisms developed above to cases of realistic photochemical air pollution, some means must be devised to convert the laboratory-derived rate constants to those suitable for atmospheric processes. Although the atmosphere over a city contains a variety of hydrocarbon species, we will attempt to characterize it by a single averaged species. The averaging will be in the sense of reactivity—*i.e.*, the hydrocarbon's susceptibility to oxidative attack, its ability to produce oxidants, or its likelihood to lead to undesirable biological effects. To do this we will relate the reactivity of pollutant species to the pure species observed in the laboratory.

When a deeper understanding of hydrocarbon synergism is attained, we can model the interactions by including the key reactions; however, an interim approach has been adopted to scale hydrocarbon rate constants according to observed reactivities. Gas chromatographic measurements made by Scott Research Laboratories (52, 53) give the input information used in the analysis. Air samples were analyzed at a central basin station and at a northeastern basin station for two smog seasons.

Before interpreting the results of the analysis, we examine the reactivity scales and their application. Previously, reactivity ratings have been assigned to the individual hydrocarbon compounds occurring in photochemical air pollution (39, 54, 55, 56). Various classification criteria have evolved on the basis of different observable effects. Reactivity response numbers represent the relative extent to which each effect occurs; some of the effects are hydrocarbon consumption rate, nitric oxide consumption rate, nitrogen dioxide peaking time, product yields, and biological responses. Also, several experimental studies on distributions of specific hydrocarbons in polluted atmospheres have been previously reported (57, 58, 59, 60, 61). Each of these studies analyzed data that are limited either in number of samples or number of compounds when compared with the body of data presented for analysis in the present program. Consequently, our data base permits us to proceed in drawing general conclusions. Table IV shows the two reactivity scales used for this analysis.

We use the reactivity groupings as a method of classifying compounds to give some insights into the processes occurring. Means and standard deviations for the values at each time of day are derived using the whole sample of days available. For the level of detail in this work, this approach gives more breadth to the results than one of focusing

Table IV. Hydrocarbon Reactivity Scales (55)

| | | Reactivity Response | |
| | | Hydrocarbon Consumption | Product Formation and Biological Effects |
Group No.	Class of Compounds		
1	C_1–C_5 paraffins, acetylene, benzene	0	0
2	C_6 + paraffins	1	1
3	Toluene and other monoalkylbenzenes	3	3
4	Ethylene	4	4
5	Dialkyl- and trialkyl-benzenes, diolefins	8	6
6	1-alkenes	17	7
7	Internally double-bonded olefins	100	8

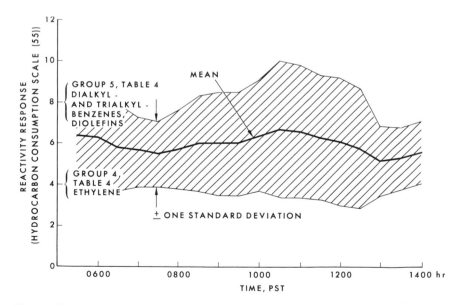

Figure 10.   Average reactivity history for Commerce, 1969: hydrocarbon consumption scale (from Altshuller)

attention on a limited number of compounds or on a limited number of days.

Figures 10 and 11 show reactivity histories at two stations averaged over all 1969 data days in the Scott program (53), using the hydrocarbon consumption scale (55). Our interest in this scale is directed mainly to the modeling problem. In these graphs and in succeeding ones we have aggregated the hydrocarbons of non-zero reactivity response and have determined the mole fraction within each group number. The reactivity for any particular time is found by summing the products of the reactivity response of each group and the mole fraction in the group. Although summation of these products may not be a precise way of characterizing the mixture, we have used it for our analysis because generalized algorithms are not yet available for computing overall reactivity.

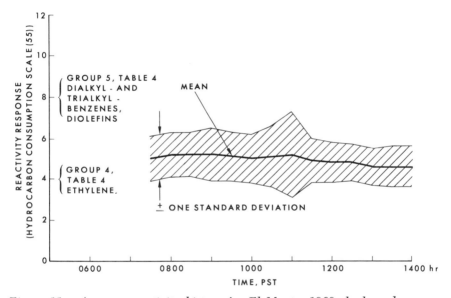

Figure 11.   Average reactivity history for El Monte, 1969: hydrocarbon consumption scale (from Altshuller)

Examining the results in Figures 10 and 11, we note that the reactivity at Commerce picks up in the hours following the morning traffic but decreases in the early afternoon hours as the photochemical fractionation overtakes the input of new hydrocarbons. All changes are small with a mean reactivity value of about six. Because of its remoteness from the strong sources, El Monte has a lower and flatter reactivity mean of about five, the scatter at El Monte is far less than that at Commerce.

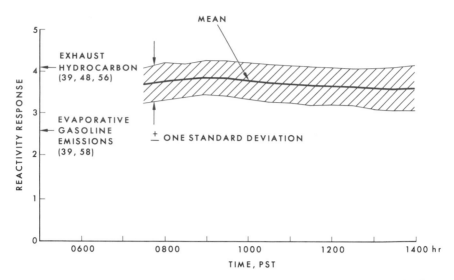

Figure 12.   Average reactivity history for Commerce, 1969: biological effects and product formation scale (from Altshuller)

See *Table II for reactivity responses of compound groups*

For adapting a simple kinetic model to the atmosphere, the behavior of the average reactivity is advantageous. The flatness of the histories and the modest scatters about the mean suggest that our propylene validation cases be scaled down by factors of two or three in the hydrocarbon rate constants. This adjustment arises because propylene is in Group 6 on Table IV showing a hydrocarbon consumption response of 17.

Besides kinetic parameters, we should check the influences of various hydrocarbons on receptors to see if they show the same type of averaged behavior. More significant from an abatement point of view are reactivity scales based on biological effects and on product formation; Figures 12 and 13 show reactivity histories using such scales resulting from Altshuller's work (55); the reactivity responses of exhaust hydrocarbons and evaporative gasoline emissions are indicated in the figures for comparison. Only slight variations are noted throughout the daytime hours. Commerce begins at its highest point at 0600 hours PST and decreases only slightly thereafter. As might be expected from windborne transport from the west, El Monte experiences its peak at about 0900 hours PST after which time it decreases slightly to about the same values as those at Commerce in the early afternoon. Both curves move only narrowly between reactivity responses of 3.4 and 3.9. Slightly higher peak reactivities are observed at El Monte than at Commerce (in contrast with the hydrocarbon consumption scale results); however, the relative flatness in the average curves and the tightly grouped data that generate the curves still prevail. Consequently, on the basis of hydrocarbon consumption and of biological effects, we can begin to approach the atmospheric kinetics problem with a single lumped hydrocarbon. Before incorporating this in the diffusion model, the properties of the governing equations will be examined, and an efficient numerical solution method will be described.

The Atmospheric Model and Its Numerical Solution

The Governing Equations. Having a simplified chemical scheme, we will now incorporate it in an atmospheric transport model. We have looked for approximations appropriate to air basins subject to photochemical pollution. Because of the severity of the problem and the availability of data, the Los Angeles basin area serves as the object of model qualification studies.

Probably the most characteristic causative factor in Los Angeles smog is the lid created by the stagnation of warm air subsiding on the cool marine layer. Late in the summer air flows from an anticyclone over the Pacific Ocean and compresses as it loses altitude in its outward course. Onshore movement of marine air under this outflow maintains a

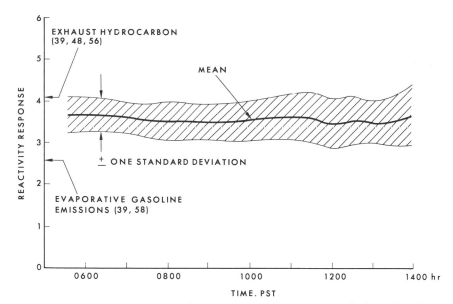

Figure 13. Average reactivity history for El Monte, 1969: biological effects and product formation scale (from Altshuller)

See *Table IV for reactivity responses of compound groups*

buoyantly stable interface at the base of the elevated inversion which is marked by a temperature increase with height. Sometimes this inversion base starts out at ground level in the morning. This condition, which is chronic and of large scale, should not be confused with the transient inversions resulting from nocturnal radiative cooling.

The regularity shown by this phenomenon holds the key to extensive reduction in atmospheric model detail. As mentioned previously, the relatively thin (marine) layer between the ground and the inversion base contains the bulk of the pollutants. The diurnal horizontal flow patterns within the marine layer are almost reproducible from day to day of the smog season. Periodic variations in height (above land) are experienced by the elevated inversion base over any given day. Some of the best detailed data illustrating these points are found in a report by Neiburger and Edinger (62). A subsequent report (63) in the same series has some graphs which associate smog events with the meteorological conditions.

The entire picture described above strongly suggests that we make a marine layer model with a prescribed internal flow field. The top of the layer is nearly as wall-like as the bottom because vertical disturbances are strongly resisted by buoyant forces. Therefore, little Reynolds stress momentum transfer occurs across the interface just as there is almost no normal transport of species (a property with which Angelenos are familiar).

For the species budget along a stream tube in the marine layer, we assume a channel-like flow with zero flux through the top wall and some specified flux from the ground. In choosing boundary conditions for the balance equations governing free radicals, we are confronted with the usual uncertainties regarding wall catalysis. Computational tests, ranging from infinite to zero catalytic efficiency for removal of oxygen atoms and RO_2, result in unimportant differences at instrument elevations of several meters because the radicals achieve steady state much faster than they diffuse over several meters.

The governing species equation is taken to be

$$\frac{\partial c}{\partial t} + u \frac{\partial c}{\partial x} = \frac{\partial}{\partial z} \left(D \frac{\partial c}{\partial z} \right) + R \tag{19}$$

where c = mass concentration

t = time

u = wind speed

x = downwind distance

z = height above ground

D = vertical turbulent diffusion coefficient

R = production rate

Mean vertical advection is suppressed by the channel-like character of the marine layer, and horizontal diffusion is relatively unimportant because of a nearly uniform distribution of emission sources. Hence these terms do not appear in Equation 19. The chemical source term is calculated from the usual rate expressions.

Since a significant period following sunrise on a high smog day has little wind, much is to be learned by omitting the second term and studying slab solutions of the heat-equation-with-sources form. As will be summarized below, this was the first approach that we followed (64) to extend laboratory data on kinetics to the atmosphere. The problem without advection (second term, Equation 19) was solved using the Crank–Nicolson technique (65). To test advection and diffusion without chemistry, we modeled carbon monoxide only using Equation 19 with $R = O$. Using Eulerian (ground-fixed) coordinates, the complete equation with the simple chemistry ran a minute of CDC 6400 time for each minute of real time that was simulated. This was considered to be unacceptable. (Slow computation resulted for the ten-cell test case because of the chemistry. It also had artificial diffusion which is an error from numerical differencing of the equation. It produces effects that resemble physical diffusion in the x-direction.)

We also use a restricted form of Equation 19 for the kinetics studies described previously. Smog chamber analyses uses just the first and last terms so that they depend on ordinary differential equations. These are solutions which describe the time-dependent behavior of a homogeneous gas mixture. We used standard Runge–Kutta techniques to solve them at the outset of the work, but as will be shown here, adaptations of Pade approximants have been used to improve computational efficiency.

The combined needs for computing efficiency and meteorological realism have been met by improvements described here; the incorporation of a classical approximation method into our numerical integration has been described above. The implementation of these computing techniques strengthens the study by allowing numerous modeling trials on laboratory and atmospheric cases. The change from Eulerian to semi-Lagrangian coordinates in the meteorological formulation is outlined below. By following air masses, we use a natural coordinate system, eliminating the artificial diffusion errors.

Application of Padé Approximants to Smog Chamber Calculations. The chemical model described previously is implemented in rate equations which must be integrated numerically. Because of the widely varying reaction rates, the equations have a characteristic known as stiffness which makes their integration difficult. For economy and accuracy, special techniques must be used to obtain a solution. Below we describe a method which uses Padé approximants (66, 67). For these ordinary differential equations, the procedure has yielded significant speed improvements over the Runge–Kutta method used previously.

For this class of problems, the implicit approach is the basic advantage of the new method. Numerical stability is thus guaranteed, and the size of the integration step is only limited by accuracy considerations. The concomitant decrease in the number of integration steps is the principal gain in computing efficiency.

The rudiments of the integration method are twofold. One involves a linearization of the nonlinear chemical term at every step; the other an approximation of the exponential term that appears as a result of the linearization operation. Thus, consider the following system of equations that describe the changes in species concentration resulting from chemical reaction:

$$\frac{dc}{dt} = R(c) \tag{20}$$

where c = vector of species concentrations

$R(c)$ = vector of chemical reaction rate functions

$$c = \begin{bmatrix} c_1 \\ c_2 \\ \cdot \\ \cdot \\ \cdot \\ c_s \end{bmatrix} \qquad R(c) = \begin{bmatrix} R_1(c) \\ R_2(c) \\ \cdot \\ \cdot \\ \cdot \\ R_s(c) \end{bmatrix}$$

s = number of chemical species

To linearize the rate vector $R(c)$, we expand it about c_o in a Taylor series. The subscript o denotes evaluation at some time $t = t_o$. Thus

$$R(c) = R(c_o) + J(c - c_o) + O(\Delta c)^2 \tag{21}$$

where

$$J = \left[\frac{\partial R_i}{\partial c_j} \right] \quad i,j = 1, 2, \ldots, s$$

is the matrix of order $(s \times s)$ of first derivatives (the Jacobian matrix) of the rate function. Substituting Equation 21 into Equation 20 yields

$$\frac{dc}{dt} = Jc + B \tag{22}$$

where $B = R(c)_o - Jc_o$. At time $t = t_o$ the quantities c_o, $R(c_o)$, and J are known. Hence, the matrix differential equation (Equation 22) is a linear equation with constant coefficients and a constant forcing function. Its formal solution is, therefore,

$$c(t) = e^{J \Delta t} [c_o + J^{-1}B] - J^{-1}B \tag{23}$$

where J^{-1} denotes the inverse (or reciprocal) of the matrix J.

An approximate solution of Equation 22 is obtained from Equation 23 by suitably approximating the matrix exponential $e^{J \Delta t}$. This is accomplished by the Padé approximants of the exponential function. These Padé approximants are rational functions of the form

$$e^{J \Delta t} \simeq \frac{P(J \Delta t)}{Q(J \Delta t)} = Q^{-1}P \tag{24}$$

where $P(J \Delta t)$ and $Q(J \Delta t)$ are matrix polynomials of degree p and q, respectively. They are defined by

$$P = \sum_{k=0}^{p} \frac{(p + q - k)\,!p!}{(p + q)\,!k!\,(p - k)\,!} (J \Delta t)^k \tag{25a}$$

$$Q = \sum_{k=0}^{q} \frac{(p + q - k)\,!q!}{(p + q)\,!k!\,(q - k)\,!} (-J \Delta t)^k \tag{25b}$$

Thus, substituting Equation 24 into Equation 23, we obtain

$$c(t) = Q^{-1}P [c_o + J^{-1}B] - J^{-1}B \tag{26}$$

The choices $p = q = 1$ and $p = q = 2$ are especially useful. The former yields a third-order integration formula, the latter a fifth-order formula. For example, for $p = q = 1$, we obtain from Equation 25:

$$P = I + 1/2\ J \Delta t \text{ and } Q = I - 1/2\ J \Delta t \tag{27}$$

where I is the identity matrix. Substituting Equation 27 into Equation 26 yields, after the appropriate manipulations,

$$c(t) = c_o + [I - 1/2\ J \Delta t]^{-1}\ \Delta t R(c_o) \tag{28}$$

Table V.   CDC 6400 Central Processor Times to Run $C_3H_6 + NO + h\nu$ Chamber Experiment Simulation

Type of Run	Runge–Kutta	Padé Approximant
Rapidly changing free radical concentration	2400 seconds	122 seconds
Typical run for 150 minutes real time	250 seconds	50 seconds
Run which diverged for Runge–Kutta	267 seconds (diverged at 80 minutes, real time)	50 seconds (ran to completion, 150 minutes real time)

Equation 28 has a form that is especially well-suited for digital computation. The process requires at every time step: (1) the evaluation of $R(c)$ and J and (2) the inversion of the matrix Q. The latter is by far the more computationally costly of the two requirements. In our case the number of species determines the size of the matrix to be inverted. Since this number is not large in our model, the matrix inversion poses no special problems; fortunately, the number of species is usually smaller than the number of reactions

Table V shows examples of the gains obtained using the new computational scheme. The typical real-time/computer-time ratio was increased from 36/1 to 180/1. Perhaps, more significant is the fact that the Padé method has allowed us to obtain acceptable solutions in situations where Runge–Kutta either failed to converge or produced spurious solutions. One such instance is the integration of full differential equations for the free-radical species. The Padé method successfully computed solutions without algebraic substitution of stationary-state assumptions whereas Runge–Kutta failed to produce any solution.

As mentioned previously, the step size must still be controlled to preserve accuracy. Referring to Equation 21, we can see that the difference $\Delta c = c - c_o$ determines the order of the truncation error of the linearization operation. The step size must be constrained so that Δc will remain within some bound specified by the user. Our codes contain a variable-step-size feature which ensures that Δc will stay within the prescribed bounds.

In our chemical model the reactions are of second order at most. Thus, the full expansion of the rate function, $R(c)$, contains no more than three terms per reaction. In the linear Taylor approximation only one term has been dropped from the expansion.

Application of Padé Approximants to the Atmospheric Model Equation. We now describe the use of Padé techniques to solve the diffusion equation with chemical reactions. The equation in question is referred to a moving air mass. Use of the Lagrangian approach (The Lagrangian approach we use is a moving air parcel with vertical nonuniformity resulting from finite rate diffusion. It was introduced by schematic description in Figure 1.) eliminates the second term in Equation 19 which is reduced to

$$\frac{\partial c}{\partial t} = \frac{\partial}{\partial z}\left(D\,\frac{\partial c}{\partial z}\right) + R(c) \tag{29}$$

where D is the vertical diffusion coefficient and c and $R(c)$ have been previously defined. Applying the "method of lines" to Equation 29, the partial differential equation is transformed into an ordinary differential equation for a line which describes the concentration changes for a fluid mass moving along the line. The equation for the ith line is derived from Equation 29 to give

$$\frac{dc_i}{dt} = \frac{D}{(\Delta z)^2}\,[\,c_{i-1} - 2c_i + c_{i+1}] + R_i \tag{30}$$

where $R_i = R(c_i)$, $i = 1, 2, \ldots, M$, where M is the number of mesh points from the ground up to the mixing depth. The symbol c_i defines a vector of s species at the ith point in space. Similarly, R_i defines a vector of s rate functions at the ith mesh point. Thus,

$$c_i = \begin{bmatrix} c_{1i} \\ c_{2i} \\ \cdot \\ \cdot \\ \cdot \\ c_{si} \end{bmatrix} \qquad\qquad R_i = \begin{bmatrix} R_{1i} \\ R_{2i} \\ \cdot \\ \cdot \\ \cdot \\ R_{si} \end{bmatrix}$$

To solve Equation 30 by the Padé method, R_i is linearized as in Equation 21 to yield

$$R_i = R_{io} + J_i\,(c_i - c_{io}) + O(\Delta c_i)^2 \tag{31}$$

where the zero subscript again denotes evaluation at time $t = t_o$, and J_i is the Jacobian matrix of the rate function at the ith spatial mesh point. Substituting Equation 31 into Equation 30 yields

$$\frac{dc_i}{dt} = \lambda\,[c_{i-1} - 2c_i + c_{i+1}] + J_i\,(c_i - c_{io}) + R_{io} \tag{32}$$

where $\lambda = D/(\Delta z)^2$. Writing Equation 32 more compactly, we obtain after collecting terms

$$\frac{dC}{dt} = (\lambda A + J) C + R_o - JC_o$$

$$\frac{dC}{dt} = HC + B \tag{33}$$

where $H = \lambda A + J$, $B = R_o - JC_o$, and

$$
C = \begin{bmatrix} c_1 \\ c_2 \\ \cdot \\ \cdot \\ \cdot \\ c_M \end{bmatrix}
\qquad
A = \begin{bmatrix}
-2I & I & 0 & \cdot & \cdot & \cdot & 0 \\
I & -2I & I & \cdot & \cdot & \cdot & 0 \\
0 & I & -2I & I & \cdot & \cdot & 0 \\
\cdot & & & & & & \cdot \\
\cdot & & & & & & \cdot \\
\cdot & & & & & & \cdot \\
\cdot & \cdot & \cdot & \cdot & \cdot & I & -2I
\end{bmatrix}
$$

$$
J = \begin{bmatrix}
J_1 & & & & \\
 & J_2 & & & \bigcirc \\
 & & \cdot & & \\
 & & & \cdot & \\
 \bigcirc & & & & \cdot \\
 & & & & & J_M
\end{bmatrix}
\qquad
R_o = \begin{bmatrix} R_{1o} \\ R_{2o} \\ \cdot \\ \cdot \\ \cdot \\ R_{Mo} \end{bmatrix}
$$

Equation 33 now has the same form as Equation 22, and its formal solution is thus analogous to Equation 23. Thus,

$$C(t) = e^{H\Delta t} [C_o + H^{-1}B] - H^{-1}B \tag{34}$$

Approximating the exponential by

$$e^{H\Delta t} = \frac{I + 1/2\ H\Delta t}{I - 1/2\ H\Delta t} = \frac{P}{Q} = Q^{-1}P$$

and substituting into Equation 34 and simplifying, we obtain

$$C = C_o + Q^{-1}G_o\,\Delta t \tag{35}$$

where

$$G_o = \lambda AC_o + R_o$$

Any boundary conditions are implicitly contained in Equation 35.

Equation 35, which is the analog of Equation 28, may appear to be simple, but the inversion of Q poses some problems arising from its dimensions. H is an $(M \times M)$ matrix, each one of whose entries is an $(s \times s)$ matrix. Thus, Q is an $(sM \times sM)$ matrix, and its storage as well as its inversion may be difficult. For example, if we have ten species ($s = 10$) and ten space grid points ($M = 10$), then Q is (100×100) and requires rather long computing times as well as large amounts of core storage. However, we take advantage of the structure of Q to develop an algorithm that reduces the problem to the inversion of M matrices of size ($s \times s$).

Since J is a diagonal matrix and A is tridiagonal, then H is tridiagonal. Moreover, the off-diagonal entries of H are identity matrices. Forming Q, we have

$$Q = I - 1/2 \, H\Delta t$$

and Q is apparently tridiagonal, its off-diagonal entries being identity matrices also. Thus, only the M diagonal entries of Q need be stored, thereby reducing memory requirements to Ms^2 cells compared with the maximum of M^2s^2. Writing Equation 35 in the form,

$$QC = QC_o + G_o\Delta t$$

we note that the tridiagonal property of Q allows us to solve for C without explicitly inverting Q. This simplification is afforded by a block form of Gaussian elimination (65) which requires the inversion of M matrices of size ($s \times s$); this is preferable to inverting the full ($sM \times sM$) matrix. We can also take advantage of the fact that the off-diagonal elements of Q are identity matrices to reduce the amount of computation.

Because series are truncated in these approximations, some remarks about the resultant errors are necessary. The order of the truncation error of the solution is $O(\Delta t)^3$. This occurs because the approximation of the exponential is

$$e^{H\Delta t} \simeq \frac{I + 1/2 \, H\Delta t}{I - 1/2 \, H\Delta t}$$

Also, as in the solution of the chemical rate equations, a linearization error of order $O(\Delta C)^2$ appears in the approximate solution. Thus, the same precautions that were taken to preserve accuracy in solving the ordinary chemical rate equations are required in the solution of the atmospheric model equation.

The new scheme has yielded a better than fourfold increase in the real-time/computer-time ratio—*i.e.*, from 7/1 to 30/1. A more detailed description is given in a separate report by Martinez (68).

Meteorological Reformulation of Model. Long computer runs and unacceptable error propagation hampered the operation of the earlier time-dependent advection and kinetics (TADKIN) code (64). Its application was limited to an analysis of carbon monoxide to examine the influence of advection added to that of vertical diffusion. At the beginning of the present work, these severe limitations required a reformulation of the advective description before any atmospheric runs were attempted. Combined with the mathematical advances just described, the semi-Lagrangian embodiment of the diffusion with kinetics (DIFKIN) code extends the application of photochemical diffusion modeling to many more cases than previously thought practicable.

Following a center-of-mass fixed coordinate system tied to an air mass, we use intrinsic coordinates to avoid artificial diffusion in the horizontal direction. Physical diffusion, therefore, is distinct and identifiable because the moving control volume can be allowed to undergo mass exchange with a neighboring air mass in a prescribed fashion. The question of horizontal spatial resolution is answered by a selection of source grid size, and the vertical resolution is set by the choice of the interval size in the z-direction.

Mixing depth or vertical extent of the pollution layer can be handled in one of two ways. In the first the vertical diffusion coefficient is assigned spatially and temporally dependent values according to the combined effects of shear and buoyancy as turbulent energy sources. The vertical extent of the grid is chosen far above the conventional mixing depth so that the pollutants are automatically confined vertically to a degree which depends on the diffusion coefficient profile. In the second way of specifying the vertical behavior, however, the vertical grid either adjusts continuously to the input values of mixing depth or assumes an average value for the inversion base altitude. We have used this second approach because it limits the field of computation to the polluted region and conserves computer time. A zero-flux boundary condition is imposed at the top of the net.

The early checkout of the model and the results in testing the improvements we attempted in the chemical, mathematical, and meteorological aspects of the problem are summarized below. Many times it seemed that the earlier simple concepts gave better results, but working with better data and more demanding theoretical descriptions than before, we uncovered some significant areas to investigate further.

The Exercise and Substantiation of the Model

Objectives of the Tests. With the development of a more complete chemical model than the previous one, the mathematical and meteorological improvements offset the potential computing slowdowns as a result of added reactions. Having described these advances in technique, we now attempt to test the model. Huntington Park is the site of the earlier tests, but El Monte is the station of interest in the recent work. El Monte is closer to the edge of the urbanized areas than Huntington Park or Commerce and is therefore likely to show larger fluctuations of pollutant concentrations. From the variability of wind directions, this is expected because the air over El Monte may have originated from either high or low pollution areas (*see* Figures 15, 16, 17, and 18). The accuracy of the initial conditions for the test calculation may be as important as the accuracy of the flux boundary conditions. Here we will examine sources of data, tests with simple chemistry and stagnant conditions, and tests with complex chemistry and advection from one station to another.

Sources of Data. INPUT INFORMATION. Except for initial profiles, the input data for the model are the meteorological conditions and the source emission conditions. Wind speed and diffusion coefficients depend on position and time. Source inventories include the flux of each primary pollutant as it depends upon location in the basin and time of day.

Empirical parameters governing atmospheric dispersal pervade the literature on this subject. Like most cases of turbulent transport, elimination of a disposable coefficient in one place leads to a reappearance of one somewhere else. The present work uses an experimentally determined turbulent diffusion coefficient, D, in Equation 19. Near the ground and near the inversion base we must assign a height (z) dependence to the diffusion coefficient.

Going up from the ground, the functional dependence assumed for D upon z allows a linear increase to some constant value. At intermediate levels D is held constant, but approaching the inversion bases, a linear decrease with height is assumed. A variation on the lower part of the profile has been to maintain the ground value up to some point before initiating the linear increase. This variation causes little change from the original trapezoidal profile. The elevation where constancy is achieved varies from 40–100 meters, depending upon meteorological conditions.

Figure 14 shows typical values of the turbulent diffusion coefficient we assumed for our model, as compared with the values either derived or observed by other investigators. At a wind speed of 1 m/s, a surface value of 30 square meters/minute is high as compared with Pasquill's (*69*) values at 200 cm elevation (even though Pasquill's values were for windy and less stable conditions), but we use it because effective roughness in an urban area far exceed those of mud flats or high grass. The assigned plateau value is slightly higher than that for the residual turbulence of free air attributed to Rossby and Montgomery (*70*) in Priestley's monograph (*71*). The decrease approaching the inversion

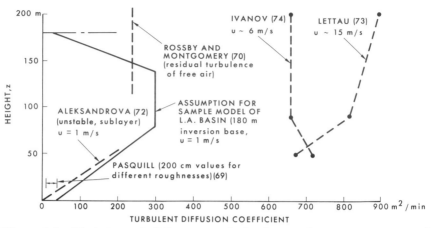

Figure 14. Variations of diffusivity with height for different average wind speeds (Ivanov and Lettau data used to determine velocity dependence)

base is introduced to reflect the wall-like inhibition of vertical turbulent motion by buoyant stability.

On Figure 14 the sublayer region corresponds closely to the values deduced from Aleksandrova's formulas and mast measurements for the precipitation of a puff released from a point source (72) (Reference 72 gives a formula $D = b_z u z$ where u is mean wind speed. For Group 2 stability (unstable layer below inversion in our case) b_z averages out to 0.064 for 18 experimental observations. The standard deviation is 0.015.) The values reported by Lettau (73) (as Austausch coefficients) were obtained from Leipzig wind profile measurements. The data of Ivanov (74) for mean square velocity and dissipation rate were used in Hanna's (75) empirical formula to obtain the values at 6 m/s. Neither the Ivanov nor the Lettau results were obtained for stable lapse conditions. The assumed values for the coefficient correspond closely to the turbulent transport observed by Craig (76), who measured water vapor profiles 30 miles offshore as warm air moved seaward across Massachusetts Bay. For light winds, values of 6–12 sq meters/minute were observed at elevations of 1–2 meters. At elevations of 60–80 meters, diffusion coefficients up to 300 sq meters/minute were found with the 6 m/s wind speeds with reduced stability. The plateau value in our profile, therefore, may be expected to vary with wind speed approximately like $50(u + 5)$ sq meters/minute where u is in m/s. This simple relationship has been used, based on the cited findings. Other sources of diffusion coefficient data (77, 78, 79) that are not plotted on Figure 14 give values which are consistent with those shown.

Wind speed, wind direction, and inversion base height are the principal meteorological inputs. Wind data are obtained from Scott

Research Laboratories (SRL) measurements (52, 53) and from other observation stations in the area. Station locations are indicated on Figure 15. Inversion data are deduced from airborne temperature measurements (52, 53) where possible and from morning soundings reported (80) by the Los Angeles County Air Pollution Control District (LAAPCD). Some wind data from the district's files have also been used.

As an example of the preparation of windfield inputs from the station data, we examine the results for one morning in the Los Angeles basin. The advection patterns are obtained by taking a weighted average of wind-station readings surrounding the point in question. This is truly a receptor-oriented modeling approach since the trajectories are constructed in hourly upwind segments from the measuring station. Reciprocal distance weighting is used because of the dominant plane flow pattern which is governed by combinations of sources, sinks, or vortices. (*See* Overview of the Combined Photochemical/Diffusion Model for a discussion of characteristic scale sizes.) In most cases the three nearest neighbor stations are included in the stepwise upwind tracing of air

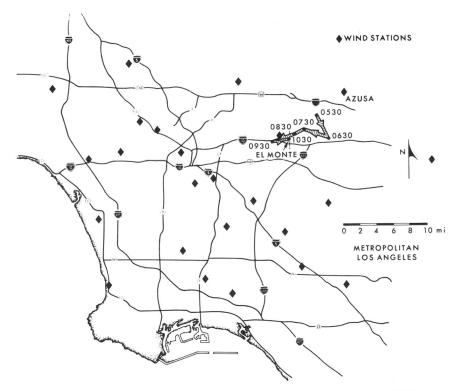

Figure 15.   Estimated trajectory of air mass arriving at El Monte, 1030 hours PST (Sept. 29, 1969)

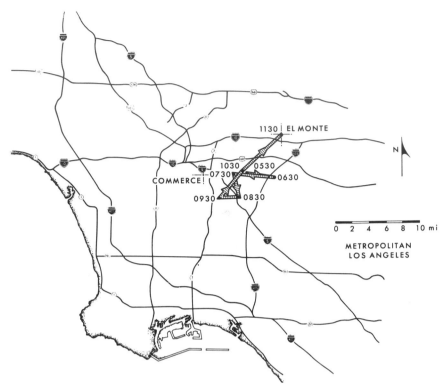

Figure 16.   Estimated trajectory of air mass arriving at El Monte, 1130 hours PST (Sept. 29, 1969)

movement, but occasionally conditions of proximity suggest including only the nearest one or two stations where coastal crossings are involved in the analysis.

Figures 15, 16, 17 and 18 show examples of the paths of air masses estimated in the manner just described. The arrival times in El Monte are used to identify each trajectory in any subsequent references in this report. The date of Sept. 29, 1969, is chosen because of the variety of data that we have available for that day. The earlier morning meandering patterns give way to the dominant onshore flows for all trajectories 0900 to 1000 hours Pacific Standard Time (PST). The meteorological formulation that we have adopted takes the time and location information from these trajectories to establish the initial conditions and the boundary conditions. Initial conditions are specified as vertical profiles of concentration and boundary conditions as time histories of surface-based pollutant emissions. These trajectories are not used in the tests of the slab model; they are used to test the moving air parcel model.

Using this approach with data from the two monitoring sites, we are able to match the computing requirements to the data base. Little justification exists at this time to apply complicated grid methods to these validation studies because we only have measurements in great detail from Commerce and El Monte. Just as overly-elaborate chemical models cannot be adequately tested in the atmosphere neither can highly complex diffusional and atmospheric formulations be checked out with our present store of field observations. The day will come when frequent high-resolution readings are made, and some of the more detailed theories can be tested.

Even more difficult than the meteorological inputs are those for surface emission fluxes of hydrocarbon, nitric oxide, and carbon monoxide species. Originally we did a simple statistical estimate on moving sources based on the California Driving Cycle for a velocity distribution function and a packing density based on a car length spacing for every 10 miles per hour speed for all the arterials and freeways in the basin. Although we arrived at surprisingly realistic values for tonnages of emission and

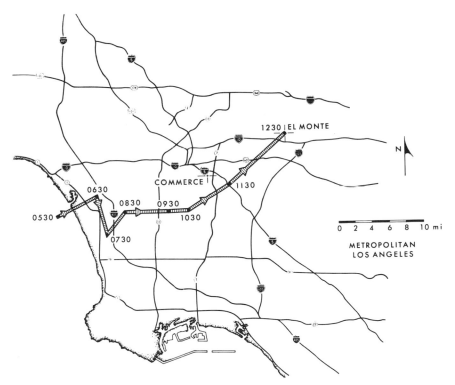

Figure 17.   Estimated trajectory of air mass arriving at El Monte, 1230 hours PST (Sept. 29, 1969)

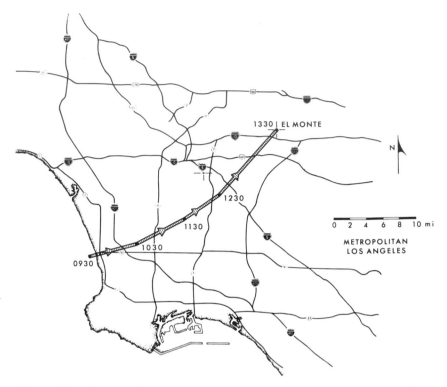

Figure 18.   Estimated trajectory of air mass arriving at El Monte, 1330 hours PST (Sept. 29, 1969)

total cars on the road, we switched to LAAPCD inventory estimates (*81*) when they became available. Table VI summarizes some of the totals from this official publication. Since tons per day are given, one can compute average surface fluxes, knowing that the basin comprises some 1250 sq miles according to the same publication.

Table VI.   Contaminants in Tons per Day from Major Sources Within Los Angeles County in 1968 (*81*)

Major Source	*Organic Gases*			*Partic-ulates*	NO_X	SO_2	CO	*Total*
	Reactivity							
	High	*Low*	*Total*					
Motor Vehicle	1255	475	1730	45	645	30	9470	11,920
Non-Motor Vehicle	200	620	820	55	305	195	225	1,610
Total	1455	1095	2550	110	950	225	9695	13,530

Additional adjustments must be made for the fact that the source emission strengths are not uniformly distributed in either time or space. To approximate the automotive sources, we used the data reported for the geographical (*see* Figure 19) and the temporal (*see* Figure 20) distributions. For nonautomotive sources averages were used as constant background emissions. The tonnage partitioning between high-reactivity and low-reactivity hydrocarbons from automotive sources roughly represents the ratio of compounds having nonzero reactivity index to those having a zero value in automotive emissions.

Although the source distribution data in Figures 19 and 20 are nearly 20 years old, they were the best available at the time of the early verification studies. The tonnages used were from 1968, and they were distributed according to Figures 19 and 20. For the more recent runs involving air trajectories, we used the source distributions developed by the group at Systems Applications, Inc. under the direction of P. M. Roth. Figures 21 and 22 were adapted from their report (83). Maps for surface street traffic and for stationary sources were also given, but Figure 21 (for freeways) typifies the data as shown on a two-mile grid. To get emissions from mileages, we used emission factors based on weighted averages of various model years according to the method of Reference 83; the weighting factors come from vehicle age distributions.

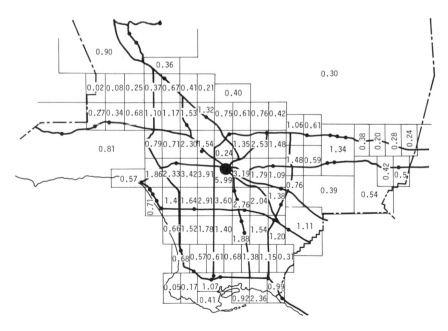

Figure 19.   Geographical traffic distribution in Los Angeles County, ca. 1951
(82)

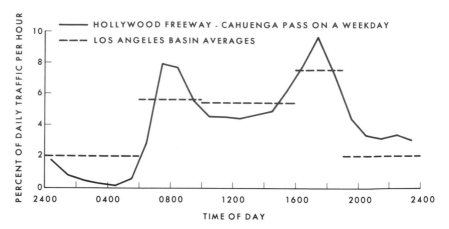

Figure 20.   Los Angeles traffic/time distributions, ca. 1951 (82)

TEST DATA INFORMATION. With the atmospheric and emission inputs, we test the model by predicting time histories of the concentration field in the marine layer. The predictions are then evaluated against a background of field data. The primary sources of these data are the SRL reports (52, 53) for the 1968–1969 smog seasons.

Because Volume I of Reference 52 contains detailed descriptions of the measurements program, it will be reviewed here only briefly. Two fixed sites, Huntington Park and El Monte, Calif. had instrumented trailers which monitored pollutant concentrations, meteorological data, and ultraviolet radiation. The sites are 12.7 miles apart and lie along a prevailing air trajectory. The great step forward in this program was the resolution of hydrocarbon concentrations into more than 100 individual species up to C_{10} with a sensitivity to levels below one part per billion using highly refined gas chromatography techniques.

Reference 53 gives descriptions and reports data for a similar program conducted in the 1969 season. The central basin station was located at Commerce, and the northeastern basin station remained at El Monte. These sites are separated by 10 miles. Great improvements in data frequency and reliability were made in the 1969 measurement program.

Limited use of other data has been made to supplement the SRL measurements. In 1967 a joint study was made by the National Center for Air Pollution Control, the California Air Resources Board Laboratory, and the Los Angeles County Air Pollution Control District. The objective of the study was directed toward improvement in detailed chemical measurement. It involved air sampling at two sites, downtown Los Angeles and Azusa. Some of the reduced data (85) have been used in the studies; the LAAPCD monitors concentrations over a nine-station network. We have used some of this information from 1968 to augment the SRL data.

The large volume of field data must be assimilated by some form of preprocessing before it is useful for testing the model. Two stages of reduction which we have used are sorting days by type characterized by peak oxidant (Days are typed by peak oxidant as follows: Type 0, both stations < 20 pphm; Type 1, either station 20–30 pphm; Type 2, either station > 30 pphm; Type 3, either station > 30 pphm *and* western station peak ⩾ eastern station peak (pphm = parts per hundred million). Western and eastern stations are taken to be Huntington Park and El Monte for SRL data, or Downtown Los Angeles and Azusa for LAAPCD data.) and then averaging data over all days of each given type. Consistent with the lumped parameter approach, we reduce the hydrocarbon

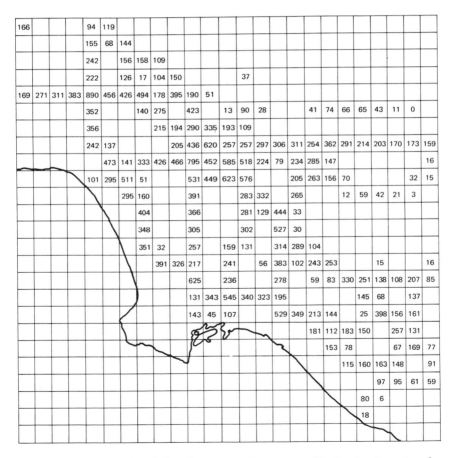

Figure 21.   Geographical distribution of freeway traffic in the Los Angeles Basin area, ca. 1968 (thousand vehicle miles per day) (83)

ensemble by grouping compounds into brackets on a reactivity scale. For a beginning only two reactivity classes have been used according to the source inventories shown in Table VI. Low reactivity hydrocarbons include methane, ethane, propane, acetylene, methylacetylene, and benzene; all others are considered to have high reactivity.

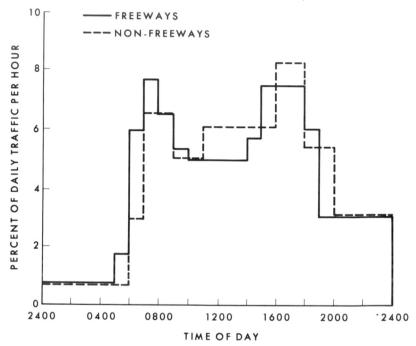

Figure 22.   Los Angeles traffic/time distributions, ca. 1968 (83)

Tests of an Early Version of the Model. In addition to studying the chemistry using chamber data, we studied atmospheric transport of inert and low reactivity species to concentrate on diffusion effects. For initial validation tests Huntington Park is superior to El Monte because of the symmetry in surrounding source distribution, the remoteness from any mountain interferences, and the regularity in the data. Our search for Type 3 days turned up October 23, 1968, as the only one with sufficient data for validation. This was used for the initial DIFKIN tests because the low wind speeds minimize the error in omitting the horizontal advection term. The simpler form of the chemical model (the one diagrammed in Figure 2) was used in these tests.

Figure 23 shows results for ground level carbon monoxide concentrations at Huntington Park. This species is practically inert so that the results are a test of the transport formulation, and the chemical source term for carbon monoxide is neglected. Using measured values to start at 0600 (Using airborne sample measurements with ground measurement for Type 2 days, we determined the early morning profiles to be well approximated by an exponential decrease with height using a 150-meter scale height.), we computed the solid curve from DIFKIN (no advection) and the dashed curve from TADKIN (input wind observations). Comparison of SRL data with output from the models shows that advection becomes significant only after the first 3 hours. Good agreement is exhibited by the models (using input data described above) with the field data. The ventilation behavior is consistent with the local wind speeds shown on an auxiliary graph on Figure 23. This approach has been applied on a small-scale grid to study the carbon monoxide pollution in the vicinity near a freeway (86).

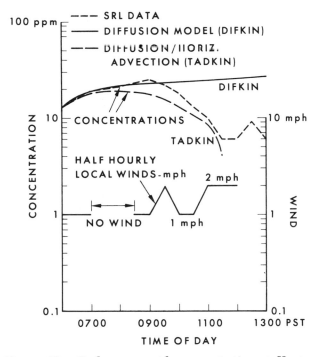

Figure 23.   Carbon monoxide concentration at Huntington Park on Type 3 day

Test results with photochemistry have been obtained for Type 2 and 3 days using the DIFKIN code. Before discussing these results, it is necessary to digress briefly to examine some special features of the data. First, consider the ratio of acetylene (C_2H_2) to total oxides of nitrogen (NO_x). In morning samples C_2H_2 is essentially unreacted, and the NO_x has not been depleted much by the chain breaking reactions.

Thus, we should expect a plateau of the C_2H_2/NO_x mole ratio characteristic of motor vehicle exhaust. Previous investigators (85) have shown
this ratio to be about four times that typifying automobile exhaust from
the California Driving Cycle. (This is observed for high-oxidant days in
1967.) These results are shown on Figure 24, suggesting that the estimates from Table VI may need modification. If we assume that nitric
oxide, the emitted pollutant, is depleted in preference to C_2H_2 (for example, by heterogeneous reactions), then we should lower the NO_x
estimates. This hypothesis is tested below by reducing the source
strengths for NO in the calculation. Other evidence leading to suspicions of the nitrogen balance is found in the ratio of CO to NO_x (87).
Carbon monoxide has long atmospheric reaction times, and since it's
origin is nearly exclusively vehicular, it too can be considered as a tracer.

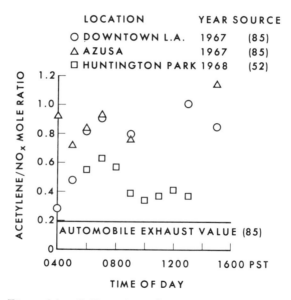

Figure 24. C_2H_2/NO_x mole ratios for high oxidant days

Figures 25 and 26 are plots of the CO/NO_x mole ratios as they vary
throughout an average day of each type. Figure 25 shows the anomaly
occurring on high oxidant days in 1968. Morning peaks of CO/NO_x ratio
seem to follow traffic curves. They exceed by large amounts the vehicular
ratio of 24 and the all-sources ratio of 16. At its maximum the average
for high-oxidant days exceeds the all-sources value by a factor of four.
This is the basis of the inconsistency between air levels and source inventories discovered in the modeling study. The physical explanation for
this inconsistency is not yet available; further diagnostic measurements
will be needed to isolate the phenomenology. Even the low-oxidant

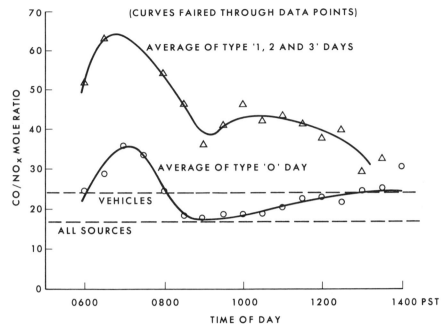

Figure 25.   CO/NO$_x$* mole ratios for Huntington Park, 1968*

curve exceeds the source ratios in the morning. It seems to follow the high-oxidant curve at a fixed interval below it until mid-morning when both curves begin to merge.

It seems reasonable to suspect that there are lower morning NO_x levels some days and that this raises peak oxidant. We will see from

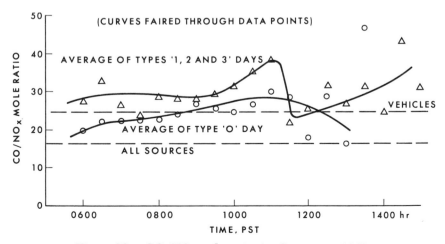

Figure 26.   CO/NO$_x$* mole ratios for Commerce, 1969*

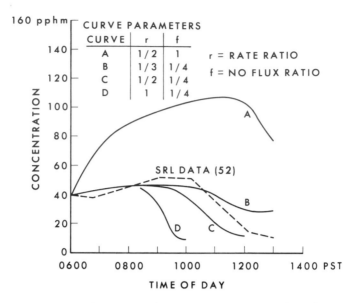

*Figure 27. (NO + NO₂)—Concentration ground level, Hunt-
ington Park on Type 3 day*

the model studies that the morning buildups of hydrocarbon and carbon
monoxide agreed well with predictions derived directly from inventory
statistics. Thus, higher HC/NO_x ratios are associated with higher peak
oxidant, as shown by the chamber experiments reported in Reference 48.

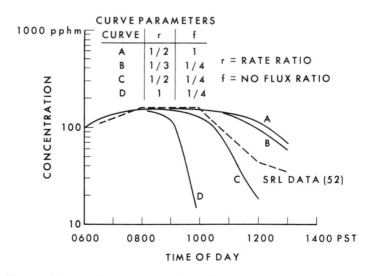

*Figure 28. High-reactivity hydrocarbons concentration on the
Type 3 day*

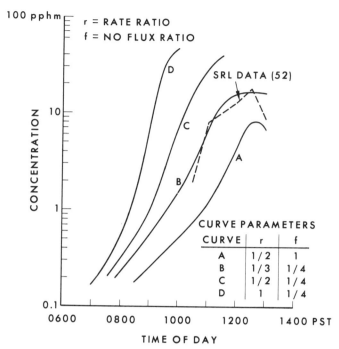

Figure 29.   Ozone concentration at Huntington Park on Type 3 day

The 1969 CO/NO$_x$ ratios do not show the strong association with peak oxidant that the 1968 data show. Figure 26 for Commerce indicates that high- and low-oxidant days have ratios only slightly higher than the emission source values. The high-oxidant days are likely to have an NO$_x$ deficit rather than a superabundance of CO.

The lack of similarity between 1968 and 1969 nitrogen oxide balances underscores the need to select a wide sample of days for model validation. Conclusions drawn from a single smog episode cannot be relied upon to establish a general abatement strategy. One suspects that sampling for a particulate nitrate may be a significant step toward resolving the problems of nitrogen oxide balances. Nitrogen dioxide oxidation processes and interaction with water vapor may form nitric acid which then leads to nitrates in solution or other condensed phases.

The initial tests of the model (DIFKIN) combining diffusion with kinetics were conducted for the Type 3 day. The rate ratio and NO$_x$ source strength ratio hypotheses were tested, as shown in Figures 27, 28, and 29. The first set of graphs illustrates most clearly the need for a downward adjustment in the nitric oxide source strength (or flux). Consistent with the results of References 85 and 87, Figures 24, 25, and 26 suggest a reduction. Using $f = 1/4$ (a fourfold strength reduction), we see a great improvement in the model predictions in the NO_x curves in Figure 27. Because of our present inability to represent the gas–liquid or gas–solid reactions mentioned above, this flux reduction is not merely a curve fitting but is likely to reflect real physical processes. For these reasons the $f = 1/4$ adjustment must be regarded as a tentative correction, pending the availability of further research data on removal processes for oxides of nitrogen. The ratio of rates is examined in these three figures by comparing curves B, C, and D. The symbol r refers to the ratio of the atmospheric-hydrocarbon oxidation rate constants to those of propylene. Evidently something between $1/2$ and $1/3$ the propylene oxidation rate constants agrees best with the observed values (52). This tends to support the hypothesis that the reactivity index goes like the oxidation rate constants in the results shown in Figures 10 and 11.

Expanding the experimental data to a larger sample, we selected Type 2 days (for which data were available during the 1968 smog season). The statistical preprocessing program derived half-hourly means

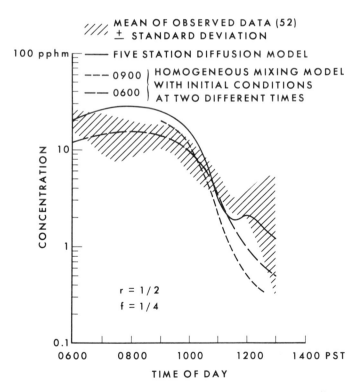

Figure 30.   Nitric oxide concentration at Huntington Park on
Type 2 days, 1968 ($f = 1/4$; $r = 1/2$)

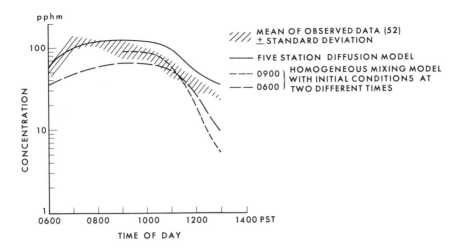

Figure 31. High reactivity hydrocarbon concentration at Huntington Park on Type 2 days, 1968 (f = 1/4; r = 1/2)

and standard deviations for this sample of days. Hatched areas on Figures 30, 31, and 32 show the mean values plus or minus one standard deviation. The DIFKIN modeling was again tested using five mesh stations, including the ground and the inversion base; moreover, a homoge-

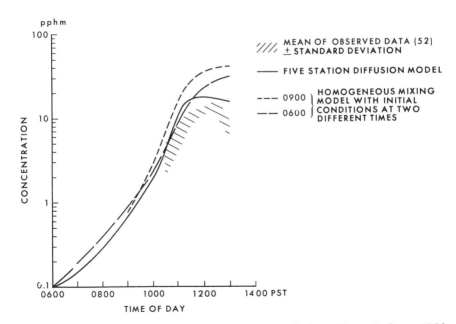

Figure 32. Ozone concentration at Huntington Park on Type 2 days, 1968 (f = 1/4; r = 1/2)

Table VII. Results of Model Sensitivity

El Monte Trajectory Arrival Time (PST)	NO			HC		
	Measured Values	(1/2) ϕNO Full ϕHC	(1/4) ϕNO (1/2) ϕHC	Measured Values	(1/2) ϕNO Full ϕHC	(1/4) ϕNO (1/2) ϕHC
1030	3.4	0.5	0.3	43	34.3	24
1130	3.2	1.5	1.5	64	87.4	70.1
1230	1.5	1.1	1.0	38	67.5	48.1

[a] Concentrations at El Monte in pphm.

neous mixing model was demonstrated. This model assumes that anything entering the marine layer from the ground is vertically mixed instantaneously. Reactions proceed according to the kinetic model derived in the section on Simplified Kinetic Mechanisms. For all of the results in Figures 30, 31, and 32, the curve parameters are $f = 1/4$ and $r = 1/2$. Average profile values at 0600 PST and at 0900 PST are used to initiate homogeneous mixing runs. Some of the precision of detail is lost in this approximation. Considering that homogeneous mixing calculations only require 5–10% of the central processor time needed by the five-station diffusion runs, they may be useful for cases where parametric effects are studied for many different conditions.

Sensitivity Studies on 1969 Trajectories with the Expanded Model. Based on the semi-Lagrangian formulation of the photochemical/diffusion model, the computed endpoint composition of the air masses depends on initial conditions, flux from the ground along the trajectories, and reaction rates. For our tests we concentrate on El Monte data because much of the polluted air there comes from somewhere else. This is believed to be a more severe test of the model than that at Huntington Park. The initial conditions are based on measurements insofar as possible. The principal initial values for the 1030 trajectory are as follows for 0730 PST (given in parts per hundred million):

$$c_{HC} = 51 \text{ pphm (6 ppm C total HC)}$$

$$c_{NO_2} = 24.5 \text{ pphm}$$

$$c_{NO} = 3.5 \text{ pphm}$$

These were interpolated from the Azusa station data (88) from the Los Angeles County Air Pollution Control District. Reference to Figure 15 shows the basis of this choice. The reactive hydrocarbon value is derived from the assumptions of 34% reactive fraction and an average carbon number of 4 for the reactive family of compounds. These are seasonal

Study Using 1969 El Monte Data[a]

	NO_2			O_3	
	(1/2)	(1/4)		(1/2)	(1/4)
Meas-	ϕNO	ϕNO	Meas-	ϕNO	ϕNO
ured	Full	(1/2)	ured	Full	(1/2)
Values	ϕHC	ϕHC	Values	ϕHC	ϕHC
16.4	20.5	14.5	13.4	32.6	35.3
21.3	58.5	48.6	23.6	28.4	24.2
10.3	49.2	34.6	24.5	30.9	24.8

[b] Nominal values of emission fluxes are denoted by ϕ, thus, $(1/2)\phi$ means that we halved the fluxes derived from inventories.

averages over the 1969 measurement months (53) at El Monte. The analysis that generated the values from gas chromatographic data is discussed elsewhere (87). Similarly the 0730 initial values for the 1130 trajectory are obtained from the Commerce data log (53). They are:

$$c_{HC} = 94.5 \text{ pphm } (9.7 \text{ ppm C})$$

$$c_{NO} = 43.9 \text{ pphm}$$

$$c_{NO_2} = 17.4 \text{ pphm}$$

The hydrocarbon conversion factors here are 41% reactive fraction and an average carbon number of 4.2.

Since the 1230 trajectory begins near the coast, we used station 76 (La Cienega Boulevard) measurements for 0630 hours (PST) initial values of nitrogen oxides. No hydrocarbon data were available from that station, and an HC/NO_x-ratio of 2 was chosen. Although this is not representative of the inventory ratio, it is representative of many morning air samples. Such an observation further reinforces the suspicion of an NO_x-deficiency on high oxidant days (87); thus, the values are

$$c_{HC} = 54 \text{ pphm}$$

$$c_{NO} = 18 \text{ pphm}$$

$$c_{NO_2} = 9 \text{ pphm}$$

The location of the 1330 trajectory origin is near no station, and the data indicate that the initial pollution cannot be neglected. Consequently, results were not obtained for the 1330 trajectory.

Ultraviolet data from Reference 53 were used to get photodissociation rates for NO_2 by calibration functions developed in the data analysis study (87). Inversion base altitudes were averaged from the airborne temperature measurements provided by Scott Research Laboratories (53).

To examine parametrically the influence of inventory levels, we altered the rate of emission for ground level. Halving the nominal NO-fluxes is suggested by the departures of the CO/NO_x-ratios and the C_2H_2/NO_x-ratios obtained for high oxidant days (87). Subsequent halving of reactive hydrocarbon and nitric oxide fluxes results in the adjustment represented by $f = 1/4$ in Figures 30, 31, and 32 but preserves the HC/NO_x-ratio. In both cases the full propylene oxidation rates were used.

Table VII summarizes the results for trajectory end points. Advancing down the table, we note an ever-growing sensitivity of the results to the flux adjustments. This occurs because the increasing wind speed lengthens the trajectories, and there is larger relative exposure of the air mass to high pollutant fluxes.

In the 1030 case, hydrocarbon and nitric oxide are low, but NO_2 and ozone are high. Changes in the fluxes have marked effects on everything but the ozone. Comparing these end-point compositions with the initial values cited above, we note a strong dependence on initial conditions accuracy for this short time scale. However, the Azusa data are available as an aid in this respect.

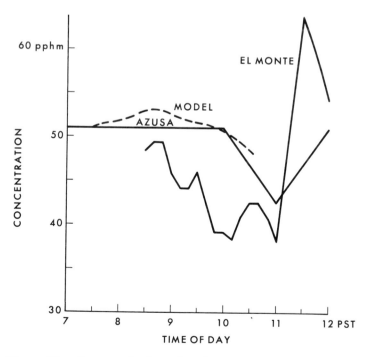

Figure 33. Reactive hydrocarbon history along the 1030 trajectory (f = 1/4; r = 1/2; ground level concentrations)

For the 1130 results, the ozone and nitric oxide come in closer to the data, but NO_2 is extremely high. This is symptomatic of the large departures from the radiation-supported quasi-equilibrium noted in analyzing the El Monte data (*87*). Although the model does not assume quasi-equilibrium among the three reactions,

$$h\nu + NO_2 \rightarrow NO + O$$

$$O + O_2 + M \rightarrow O_3 + M$$

$$O_3 + NO \rightarrow NO_2 + O_2$$

the differential equation solutions nearly always approximate it rather well.

Again with the 1230 trajectory, the ozone and nitric oxide levels match observations reasonably well, but hydrocarbon and NO_2 are both high. Despite the passage of this air from the seashore and then over extensive regions of the central basin, we note that the initial values of the primary pollutants still influence the final concentrations in a dominant manner.

History Analysis of the 1030 El Monte Trajectory. Because of the relative completeness of initial conditions that we can relate to the Azusa station, we have chosen the 1030 trajectory to discuss in some detail. Examining Table VII, we note an overabundance of ozone at 1030 and a correspondingly rapid completion of $NO \rightarrow NO_2$ conversion. Reactivity analyses (*see* Atmospheric Adaptation) and our early modeling studies suggest reduction of the oxidation rate constants. To achieve some level of comparative assessment with the previous work, we assign one-fourth the nominal NO flux and one-half the oxidation rate; thus, $f = 1/4$ and $r = 1/2$ describe the conditions as before. This time, however, we preserve the HC/NO_x-ratio as in the entries in Table VII and reduce hydrocarbon fluxes by a factor of two. This means that the difference in end-point concentrations between this case and the $1/4\phi_{NO}$, $1/2\phi_{HC}$ entries results solely from the rate constant reduction. This differs from the earlier work where hydrocarbon fluxes were not reduced.

Figure 33 shows the reactive hydrocarbon history starting at Azusa at 0730 and ending at El Monte at 1030 on September 29, 1969. Despite the sharp reduction in hydrocarbon fluxes, the calculated curve stays above the Azusa levels until 1000 hours when it begins to slope-off in the observed manner. The model output clearly bears a closer relationship to the Azusa measurements than it does to the El Monte observations. This problem is typical of the difficulties we encountered in attempting to model conditions at the eastern portion of the basin at El Monte.

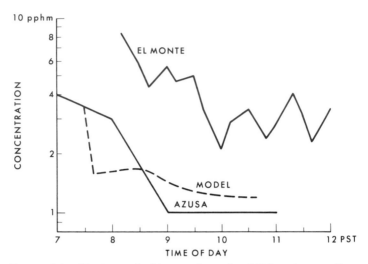

Figure 34.   Nitric oxide history along the 1030 trajectory (f = 1/4; r = 1/2; ground level concentrations)

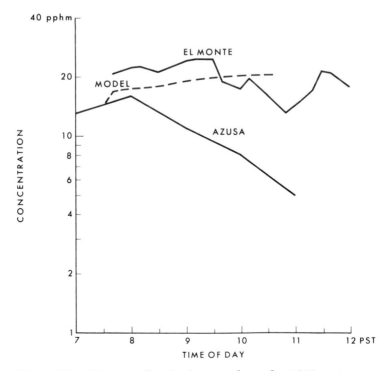

Figure 35.   Nitrogen dioxide history along the 1030 trajectory (f = 1/4; r = 1/2; ground level concentrations)

On Figure 34 we note a sharp drop from the interpolated NO-level between 0730 and 0740 hours. This reflects the previously noted failure of the data to approach quasi-equilibrium between NO, NO_2, O_3, and sunlight intensity under high-oxidant conditions. The NO-conversion seems to proceed at roughly the observed rate after the transient is absorbed in the system; however, the level ends up closer to Azusa values than to El Monte values. If we were to use 0830 El Monte concentrations as initial values, we would have a lower HC/NO-ratio and could expect still slower nitric oxide conversion rates. Thus, nitric oxide and hydrocarbon decay more like the Azusa data than the El Monte data.

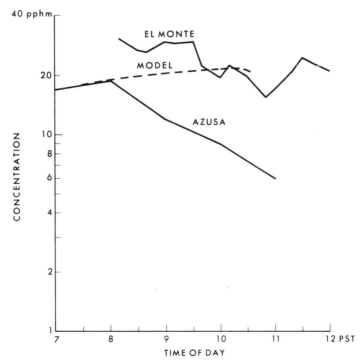

Figure 36.   Oxides of nitrogen (NO + NO_2) history along the 1030 trajectory (f = 1/4; r = 1/2; ground level concentrations)

The NO_2 behavior on Figure 35 exhibits more nearly what one would expect than do either the reactive HC or the NO. Proceeding from the end of its transient adjustment to observed sunlight intensity, the air mass gradually undergoes NO_2-transition from Azusa levels to El Monte levels as it meanders about in the northeastern area of the basin. Nitric oxide conversion supported by increasing sunlight intensities drives the NO_2

upward. The ground-based sources continue to feed in NO as it reacts and diffuses upward.

Summing NO and NO_2, we get excellent total balance behavior for these oxides of nitrogen. Figure 36 indicates that slopes and magnitudes are represented rather well in the gradual change of composition moving from initial to final conditions over the three-hour period. Apparently the dominant effect of the good behavior of the NO_2 brings the balance into favorable agreement with the data. It will be recalled that this was a key test in the choice of $f = 1/4$ after conducting our original modeling study.

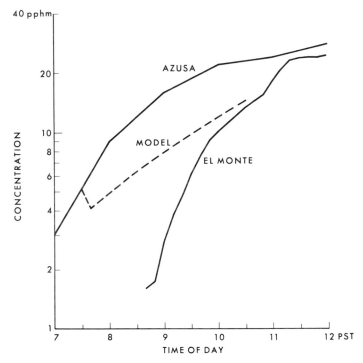

Figure 37. Ozone history along the 1030 trajectory ($f = 1/4$; $r = 1/2$; ground level concentrations)

Finally, the history of ozone concentration may be seen on Figure 37. The measurements of Azusa and El Monte show a remarkable degree of smoothness and regularity compared with the jagged curves of the species shown in the previous figures. Again, because of the rapid transient response of the $O_3/NO/NO_2$-system to the sunlight ultraviolet intensity, there is a sharp change in the first ten minutes of the model curve. As for NO_2, however, the ozone undergoes a smooth transition from its origin to its destination. Considering the usual sensitivity of ozone to competing rate determining factors, this degree of realism is gratifying.

If we applied this same set of assumptions to the other trajectories, we would get too little ozone at the end points. Table VII indicates fair agreement with $f = 1/4$ and $r = 1$ as shown in one of the columns. The hydrocarbon agreement for these other cases would be even worse than it stands if r were decreased to $1/2$. Because of low confidence in the initial values that apparently dominate, the other trajectories would likely be less than satisfactory as tests of the model.

Conclusion

Before detailing the accomplishments and recommending new directions, we make the general observation that applying and adapting the modeling tools that we possess now seems preferable to initiating effort on far more detailed chemical dispersion formulations. We have maintained a balanced posture in which the model advances and the improvements in measurements have kept pace with one another.

The chemical descriptions offered here hold the main feature of the NO_x-chain termination that controls composition effects on secondary pollutant production. Accordingly, the addition of hydroxyl radical chemistry is a first step in explaining the excess rate of hydrocarbon oxidation (over that attributable to O-atom attack). Probably one of the most uncertain aspects of the system is our inclusion of HONO formation (from $NO + NO_2$) and opposed by photodissociation. If this is an important feature of the free radical balance, we need to know the photodissociation absorption coefficients for HONO as well as some of the radical-molecule collisional reaction rates. In his task force report (89) for Project Clean Air, H. S. Johnston stresses the need to consider nitrous and nitric acids in the photochemical mechanism. This suggests a real need to monitor these compounds in the laboratory experiments and in the atmosphere. Their involvement in aerosols and surface reactions may hold the key to the apparent anomalies in the nitrogen balance. The identification of chemo-mutagenic effects of HONO is another reason that it should be investigated. The brief examination of hydrocarbon synergism shows at least one way that the radical chains are cross-linked to influence combined reaction rates. Experimental data on mixed hydrocarbons will help sort out the most likely mechanisms. Likewise, our small study on possible CO interactions indicates that suggested values of these rate constants indicate a strong influence on the propylene/nitric oxide system at concentration levels as high as 100 ppm. It may be that atmospheric hydrocarbons that are not involved in the OH-cycle are far less affected. In any event atmospheric levels of CO are much lower than 100 ppm except near traffic arteries.

Reacting to the need for extended chemical representations, we have applied Padé approximant techniques to the numerical integration algorithms. Speed increases of at least four or fivefold are obtained in the computation comparing the Padé method for the more complex chemical system with previous methods for the simpler system. For equal-sized chemical matrices, the improvements would most likely be even better. To make these advances available to the community at large, we show the mathematical steps in detail.

Besides the mathematical improvements, the atmospheric model has been adapted to a semi-Lagrangian formulation. By following selected air masses, we avoid commitments of large quantities of memory and the incursions of artificial diffusion errors. Most important, we do not end up with stacks of computer printout that relate to regions where there are no measurements. Also, predictive calculations will become more useful, but our present levels of resources and sophistication demand that effort be concentrated on verification. Only in this way can the confidence be built that is needed for applying modeling techniques to implementation planning.

Tests of the model for 1969 Los Angeles basin data turned up several useful findings. For time intervals within a diurnal scale, we must be extremely careful in selecting initial conditions. A way to overcome this need is to compute several days of real time on a continuous basis. Increasing the time scale makes sense because episodes tend to cycle through several days. The computing cost will go up more than proportionately, however, because as time scales increase so must the spatial domain. Perhaps 100 km of urban/rural influence zone might be needed in lateral expansion. Correspondingly, a larger vertical field will have to be included to account for slow transport within the inversion layer if there is one. Models that require artificial image flows above the marine layer or that assume artificially inflated background levels at the edge of the net will no longer be useful. What were boundaries will now be included in the computational field.

The sensitivity tests in the model qualification studies confirmed some of the suspicions that we expressed earlier (87)—i.e., that the quasi-equilibrium relationship between ozone and the oxides of nitrogen does not seem to be recovered in the data. The largest departures are for the highest ozone levels. Attempting to represent the physical setting consistently, we find it difficult to use the measured ultraviolet intensity to account simultaneously for the observed ozone buildup and NO-conversion simultaneously. The inconsistency even appears in the initial behavior of a modeling run as a transient induction process that rapidly adjusts concentrations to satisfy the rate equations. Unlike some models ours does not impose the quasi-equilibrium but rather solves for time history of species with O-atom, RO_2-radicals, and OH in the stationary state.

Because of the extreme dependence on initial conditions, our history analysis concentrates on an air mass with relatively well-defined concentrations at the beginning and the end of its travel. Giving it the initial values, we see the concentrations unfold as the air parcel moves through the computed simulation procedure. Because of the sensitivities discovered, the transition of oxidant species O_3 and NO_2 proceeds better than one might expect. The previously adopted biases on the NO-flux and the propylene oxidation rates were confirmed in this run having different conditions from those in Huntington Park represented by 1968 data.

Much more work remains to be done. We have so many uncertainties in the data and the model assumptions that systematic tests must be devised to isolate the individual influences of the various parameters. Our limited sensitivity study illuminates some of these serious questions that need to be answered. The anomalous pattern of NO-flux followed from our original work, through the data analysis project, and into this work. Either some serious deficiency exists in the transport formulation, or there are some rapid loss mechanisms for oxides of nitrogen that reduce the apparent emission strengths. Because of the extensive efforts directed toward NO_x-controls, priority must be given to measuring the terms of the nitrogen balance equation in an urban area. Nitric acid vapor, nitrous acid vapors, and particulate nitrates could hold the key to this question. The modeling is not precise enough to pin down the deficit, but the CO/NO_x and C_2H_2/NO_x analyses of the Los Angeles data (87) showed significant deviations on high-oxidant days. Since the same types of questions have appeared in laboratory experiments, it seems that a greater variety of nitrogen-bearing compounds should be measured in future atmospheric studies.

The plans announced for a regional air pollution study should hold many of the solutions to the problems identified here. Conversely, application of the model should give valuable feedback to such programs. The chemical mechanism should be clarified to give better values of rate constants for the controlling reactions. Perhaps the selection of which reactions to include might be altered for the air contaminant hydrocarbon mix. Mathematical techniques seem well enough advanced to match the simpler formulations of photochemical modeling schemes to the computing machinery available. The more complex schemes proposed recently will continue to shed light on questions of dispersion parameters and inventory accuracy.

The future of the mathematical modeling techniques is linked to cooperative activity between the theoretical and experimental arts in the field of air pollution. Important phenomena are yet to be added to any of the mathematical schemes. The formation of aerosol and its extinction of ultraviolet radiation has not been explicitly treated in the computations. Moreover, the whole area of heterogeneous reactions on either particulate surfaces or urban surfaces remains obscure. The reacting flow problem of mixedness and its influence on kinetics has not been reduced to an engineering procedure for calculational purposes.

Urgent needs for improved air quality management demand renewed vigor in our attack on many of the fluid dynamic and chemical kinetic areas plagued by chronic difficulties. Turbulent mixing and free radical processes are significant here as in many previous technical problems. Pragmatic perspective must guide us, however, in the development of mathematical modeling lest it become a prophecy which fulfills itself in totally abstract research activity.

Acknowledgment

The General Research Corporation portions of the work were performed pursuant to Contracts CPA 22-69-127, National Air Pollution Control Administration, Department of Health, Education and Welfare, CRC-APRAC Project No. CAPA-7-68, Coordinating Research Council, and EHSD 71-22, Air Pollution Control Office, Environmental Protection Agency.

Literature Cited

1. Barth, D. S., "Federal Motor Vehicle Emission Goals for CO, HC, and NO_x Based on Desired Air Quality Levels," *J. Air Poll. Contr. Assoc.* (August 1970) **20** (8), 519–523.
2. Callaghan, D. J., Feldstein, M., "Meeting Air Quality Standards. The Pragmatic Approach," Second International Clean Air Congress Paper AD-36H, Washington, D. C. (December 6-11, 1970).
3. Sutton, O. G., "A Theory of Eddy Diffusion in the Atmosphere," *Proc. Roy. Soc.* (1932) **A135**, pp. 143–165.
4. Sutton, O. G., "Micrometeorology," p. 34, McGraw-Hill, New York, 1953.
5. Frenkiel, F. N., "Atmospheric Pollution in Growing Communities," *Smithsonian Institute Annual Report*, pp. 269–299, 1956.
6. Pooler, F. Jr., "A Prediction Model of Mean Urban Pollution for Use with Standard Wind Roses," *Int. J. Air Water Poll.*, 1961, Vol. 4, Nos. 2/4, pp. 199–211.
7. Turner, D. B., "A Diffusion Model for an Urban Area," *J. of Appl. Met.* (February 1964) **3**, 83–91.
8. Clarke, J. F., "A Simple Diffusion Model for Calculating Point Concentrations for Multiple Sources," *J. Air Poll. Contr. Assoc.* (September 1964) **14** (9), 347–352.
9. Koogler, J. B., Sholtes, R. S., Danis, A. L., Harding, C. I., "A Multivariable Model for Atmospheric Dispersion Predictions," *J. Air Poll. Contr. Assoc.* (1967) **17**, 211–214.

10. Miller, M. E., Holzworth, George C., "An Atmospheric Diffusion Model for Metropolitan Areas," *J. Air Poll. Contr. Assoc.* (January 1967) **17** (1), 46–50.
11. Martin, D. O., Tikvart, J. A., "A General Atmospheric Diffusion Model for Estimating the Effects of One or More Sources on Air Quality," US DHEW Public Health Service, National Air Pollution Control Administration, 1968.
12. Lucas, D. H., "The Atmospheric Pollution of Cities," *Int. J. Air Pollut.* (1958) **1** (1/2), 71–86.
13. McElroy, J. L., Pooler, F., Jr., "St. Louis Dispersion Study, Volume II— Analysis," USDHEW Public Health Service, National Air Pollution Control Administration, December 1968.
14. Turner, D. B., "Workbook of Atmospheric Dispersion Estimates," US DHEW Public Health Service, National Center for Air Pollution Control, 1967.
15. Smith, M. E., "Chemical Reactions in the Lower and Upper Atmosphere," pp. 273–286, Stanford Research Institute (April 1961).
16. Wanta, R. C., "Meteorology and Air Pollution," *Air Pollution* (1968), Chpt. 7, 1, A. C. Stern, Ed., Academic, New York.
17. Lamb, R. G., "An Air Pollution Model of Los Angeles," University of California Los Angeles, M.S. Thesis, 1968.
18. Ludwig, F. L., Johnson, W. B., Moon, A. E., Mancuso, R. L., "A Practical, Multipurpose Urban Diffusion Model for Carbon Monoxide," Stanford Research Institute Report, September 1970.
19. Roberts, J. J., Croke, E. J., Kennedy, A. S., Norco, J. E., Conley, L. A., "A Multiple-Source Urban Atmospheric Dispersion Model," Argonne National Laboratory Report ANL/ES-CC-007, May 1970.
20. Karplus, W. J., Bekey, G. A., Pekrul, P. J., "Atmospheric Diffusion of Air Pollutants," *Ind. Eng. Chem.* (1958) **50**, 1657–1660.
21. Friedlander, S. K., Seinfeld, J. H., "A Dynamic Model of Photochemical Smog," *Environ. Sci. Technol.* (1969) **3** (11), 1175–1182.
22. Calvert, S., "A Simulation Model for Photochemical Smog," *Calif. Air Environ.* (July–September 1969) **1** (3), p. 1.
23. Wayne, L., Danchick, R., Weisburd, M., Kokin, A., Stein, A., "Modeling Photochemical Smog on a Computer for Decision-Making," *Ann. Meet. Air Pollut. Contr. Assoc.*, *63rd*, St. Louis (June 14-18, 1970).
24. Westburg, K., Cohen, N., "The Chemical Kinetics of Photochemical Smog as Analyzed by Computer," American Institute of Aeronautics and Astronautics 3rd Fluid and Plasma Dynamics Conference, Los Angeles (June 29–July 1, 1970).
25. Sklarew, R. C., "A New Approach: The Grid Model of Urban Air Pollution," *Ann. Meet. Air Pollut. Contr. Assoc.*, *63rd*, St. Louis (June 14-18, 1970).
26. Seinfeld, J. H., Roth, P., Raynolds, S. D., Hecht, T. A., "Simulation of the Los Angeles Basin," *161st Natl. Meet. Amer. Chem. Soc.*, Los Angeles (March 28–April 2, 1971).
27. Eschenroeder, A. Q., "Validation of Simplified Kinetics for Photochemical Smog Modeling," General Research Corporation IMR-1096, September 1969.
28. Leighton, P. A., "Photochemistry of Air Pollution," Academic, New York, 1961.
29. Saltzman, B. E., "Kinetic Studies of Formation of Atmospheric Oxidants," *Ind. Eng. Chem.* (April 1958), **50** (4), 677–682.
30. Wayne, L. G., "On the Mechanism of Photo-Oxidation in Smog," *Archives Environ. Health* (1963) **7** (8), 113–123.
31. Altshuller, A. P., Bufalini, J. J., "Photochemical Aspects of Air Pollution: A Review," *Photochem. Photobiol.* (1965) **4**, 97–146.

32. Haagen-Smit, A. J., Wayne, L. G., "Atmospheric Reactions and Scavenging," *Air Pollution* pp. 149–186, Vol. I, A. C. Stern, Ed., Academic, New York, 2nd ed., 1968.
33. Stephens, E. R., "Chemistry of Atmospheric Oxidants," *J. Air Pollut. Contr. Assoc.* (1969) **19**, 181–185.
34. Tuesday, C. S., "The Atmospheric Photo-oxidation of Trans-Butene-2 and Nitric Oxide," General Motors Research Laboratories GMR-332, 1961.
35. Bufalini, J. J., Brubaker, K. L., "Photo-oxidation of Formaldehyde at Low Partial Pressures," Symposium for Chemical Reactions in Urban Atmospheres, GM Research Laboratories, Warren (October 6-7, 1969).
36. Altshuller, A. P., DHEW, private communication (September 1969).
37. Westberg, K., Aerospace Corporation, private communication.
38. DASA Reaction Rate Handbook, DASA 1948 (October 1967).
39. Caplan, J. D., "Smog Chemistry Points the Way to Rational Vehicle Emission Control," *Vehicle Emissions II: SAE Progr. Technol.* (1966) **12**, 20–31.
40. Altshuller, A. P., Kopczynski, S. L., Lonneman, W. A., Becker, F. L., Slater, R., "Chemical Aspects of the Photo-oxidation of the Propylene-Nitrogen Oxide System," *Environ. Sci. Technol.* (1967) **1**, 899–914.
41. Kopczynski, S. L., unpublished data (October 1969).
42. Schuck, E. A., Doyle, G. J., "Photooxidation of Hydrocarbons in Mixtures Containing Oxides of Nitrogen and Sulfur Dioxide," Air Pollution Foundation Report No. 29, San Marino, California, October 1959.
43. Stedman, D. H., Morris, Jr., E. D., Daby, E. E., Niki, H., Weinstock, B., "The Role of OH Radicals in Photochemical Smog," American Chemical Society Division of Water Air and Waste Chemistry, Chicago (Sept. 13-18, 1970).
44. Holmes, J. R., Sanchez, A. D., Bockian, A. H., "Atmospheric Photochemistry: Some Factors Affecting the Conversion of NO to NO_2," Pacific Conference on Chemistry and Spectroscopy, San Francisco (October 6-9, 1970).
45. Altshuller, A. P., Bufalini, J. J., "Photochemical Aspects of Air Pollution: A Review," *Environ. Sci. Technol.* (1971) **5** (5), 39–64.
46. Seinfeld, J. H., private communication (June 2, 1970).
47. Altshuller, A. P., Kopczynski, S. L., Wilson, D., Lonneman, W., Sutterfield, F. D., "Photochemical Reactivities of n-Butane and Other Paraffinic Hydrocarbons," *J. Air Pollut. Contr. Assoc.* (1969) **19** (10), pp. 787–790.
48. Korth, M. W., Rose, Jr., A. H., Stahman, R. C., "Effects of Hydrocarbon to Oxides of Nitrogen Ratios on Irradiated Auto Exhaust, Part I," *J. Air Pollut. Contr. Assoc.* (1964) **14** (5), 168–175.
49. Agnew, W. G., "Automotive Air Pollution Research," *Proc. Roy. Soc., Series A* (1968) **307**, 153–181.
50. Heicklen, J., Westberg, K., Cohen, N., "The Conversion of NO to NO_2 in Polluted Atmospheres," Pennsylvania State University Center for Air Environment Studies Publication 115-69 (July 1969).
51. Westberg, K., Cohen, N., Wilson, K. W., "Carbon Monoxide: Its Role in Photochemical Smog Formation," *Science* (1971) **171**, 1013–1015.
52. "Final Report on Phase I, Atmospheric Reaction Studies in the Los Angeles Basin," Vols. I and II, Scott Research Laboratories (June 30, 1969).
53. "Final Report, 1969 Atmospheric Reaction Studies in the Los Angeles Basin," Vols. I-IV, Scott Research Laboratories (February 1970).
54. Stephens, E. R., Scott, W. E., "Relative Reactivity of Various Hydrocarbons in Polluted Atmospheres," *Proc. Amer. Petrol. Inst.* (1962) **42 (III)**, 665.
55. Altshuller, A. P., "An Evaluation of Techniques for the Determination of the Photochemical Reactivity of Organic Emissions," *J. APCA* (1966) **16** (5), 257–260.

56. Bonamassa, F., Wong-Woo, H., "Composition and Reactivity of Exhaust Hydrocarbons from 1966 California Cars," *Natl. Meet. Amer. Chem. Soc.*, *152nd*, New York (Sept. 11-16, 1966).
57. Neligan, R. E., "Hydrocarbons in the Los Angeles Atmosphere," *Archives of Environ. Health* (1962) **5** (12), 581-591.
58. Stephens, E. R., Burleson, F. R., "Analysis of the Atmosphere for Light Hydrocarbons," *J. Air Pollut. Contr. Assoc.* (1967) **17** (3), 147-153.
59. Gordon, R., Mayrsohn, H., Ingels, R., "C_2-C_5 Hydrocarbons in the Los Angeles Atmosphere," *Environ. Sci. Technol.* (1968) **2**, 1117-1120.
60. Altshuller, A. P., Kopczynski, S. L., Lonneman, W. A., Sutterfield, F. D., "A Technique for Measuring Photochemical Reactions in Atmospheric Samples," *Environ. Sci. Technol.* (1970) **4** (6), 503-506.
61. Stephens, E. R., Burleson, F. R., "Distribution of Light Hydrocarbons in Ambient Air," *Ann. Meet. Air Pollut. Contr. Assoc.*, 62nd (1969).
62. Neiberger, N., Edinger, J., "Meteorology of the Los Angeles Basin with Particular Respect to the Smog Problem," Air Pollution Foundation Report No. 1 (April 1954).
63. Renzetti, N. (Ed.), "An Aerometric Survey of the Los Angeles Basin August-November 1954," Air Pollution Foundation Report No. 9 (July 1955).
64. Eschenroeder, A. Q., Martinez, J. R., Mathematical Modeling of Photochemical Smog, General Research Corporation IMR-1210, December 1969. (Also AIAA paper 70-116, 8th Aerospace Sciences Meeting, New York, January 19-21, 1970.)
65. Ames, W. F., "Nonlinear Partial Differential Equations in Engineering," pp. 341-342, Academic, New York, 1965.
66. Varga, R. S., "On Higher Order Stable Implicit Methods for Solving Parabolic Partial Differential Equations," *J. Math. Phys.* (1961) **40**, 220-231.
67. Magnus, D., Schecter, H., "Analysis and Application of the Padé Approximation for the Integration of Chemical Kinetic Equations," Technical Report 642, General Applied Science Laboratories, Inc., 1967.
68. Martinez, J. R., "Improving the Efficiency in Atmospheric Reaction Modeling Calculations," General Research Corporation IMR-1291, April 1971. (Also Air Pollution Control Association Paper 71-138, 64th National Meeting, Atlantic City, June 27-July 1, 1971.)
69. Pasquill, F., "Atmospheric Diffusion," p. 72, Van Nostrand, London, 1962.
70. Rossby, C. G., Montgomery, R. B., "The Layer of Frictional Influence in Wind and Ocean Currents," *Phys. Oceanogr. Meteorol.* (1935) **3**, No. (3).
71. Priestley, C. H. B., "Turbulent Transfer in the Lower Atmosphere," pp. 33-38, University of Chicago (1969).
72. Aleksandrova, A. K., Byzova, N. L., Mashkova, G. B., "Investigation of the Bottom 300-Meter Layer of the Atmosphere," pp. 1-12, N. L. Byzova, Ed. (translated from the Russian by Israel Program for Scientific Translation), Jerusalem, 1965.
73. Lettau, H., "A Reexamination of the 'Leipzig Wind Profile' Considering Some Relations Between Wind and Turbulence in the Frictional Layer," *Tellus* (1950) **2**, 189-200.
74. Ivanov, V. N., "Investigation of the Bottom 300-Meter Layer of the Atmosphere," pp. 36-42, N. L. Byzova, Ed. (translated from the Russian by Israel Program for Scientific Translation), Jerusalem, 1965.
75. Hanna, S. R., "A Method for Estimating Vertical Eddy Transport in the Planetary Boundary Layer Using Characteristics of the Vertical Velocity Spectrum," *J. Atmospheric Sci.* (1968) **25** (6), 1026-1033.
76. Craig, R. A., "Vertical Eddy Transfer of Heat and Water Vapour in Stable Air," *J. Meteorol.* (1949) **6**, 122-133.

77. Wu, S. S., "A Study of Heat Transfer Coefficients in the Lowest 400 Meters of the Atmosphere," *J. Geophys. Res.* (1965) **70**, (8), 1801–1807.
78. Hosler, C. R., "Vertical Diffusivity from Radon Profiles," *J. Geophys. Res.* (1969) **74** (28), 7018–7026.
79. De Zorzo, G., "Applicazione della Equazione di Diffusione allo Studio della Bassa Atmosfera," *Riv. Meteorologica Aeronautica* (1970) **30** (1), 3–28.
80. Monthly Reports of Meteorology, Air Pollution Effects and Contaminant Maxima, Air Pollution Control District, County of Los Angeles.
81. Lemke, E. E., Thomas, G., Zwiacher, W. E. (Eds.), "Profile of Air Pollution Control in Los Angeles County," APCS, Los Angeles County, January 1969, p. 3.
82. Larson, G. P., Chipman, J. C., Kauper, E. K., "Distribution and Effects of Automotive Exhaust Gases in Los Angeles," *Vehicle Emissions I. SAE Progress in Technology Series* (1964) **6**, 7–16.
83. Roberts, P. J. W., Roth, P. M., Nelson, C. L., "Contaminant Emissions in the Los Angeles Basin—Their Sources, Rates and Distribution," Systems Applications, Inc. Report 71SAI-6, March 1971.
84. Eschenroeder, A. Q., "Some Preliminary Air Pollution Estimates for the Santa Barbara Region, 1970–1990," General Research Corporation IMR-1278, March 1970.
85. Gordon, R. J., Mayrsohn, H., Ingels, R. M., "C_2–C_5 Hydrocarbons in the Los Angeles Atmosphere," *Environ. Sci. Technol.* (1968) **2**, 1117–1120.
86. Eschenroeder, A. Q., "An Approach for Modeling the Effects of Carbon Monoxide on the Urban Freeway User," General Research Corporation IMR-1259, January 1970.
87. Eschenroeder, A. Q., Martinez, J. R., "Analysis of Los Angeles Atmospheric Reaction Data from 1968 and 1969," General Research Corporation CR-1-170, July 1970.
88. Monitoring Data Sheets from Los Angeles County Air Pollution Control District, 1969.
89. Johnston, H., "Reactions in the Atmosphere," p. 3, *Project Clean Air Task Force Assessments*, September 1, 1970, Vol. 4, Task Force No. 7, Section 3, University of California.

RECEIVED June 10, 1971.

Abatement Strategy for Photochemical Smog

A. J. HAAGEN-SMIT

Division of Biology, California Institute of Technology, Pasadena, Calif. 91109

Photochemical air pollution caused by sunlight acting on mixtures of oxides of nitrogen and organic materials concerns large metropolitan areas. An orderly control program identifies the nature of the problem, followed by information on ambient air, emission standards, and a plan of action. The overall responsibility for controlling photochemical smog has moved from local to state to federal authorities. In some seriously afflicted areas technological control is not enough to meet clean air standards. Here control strategy must prevent rather than cure. This strategy will affect the social and economic structure of society and will enact new stringent air pollution laws. Gradual developing and adapting of control strategy to environmental and technological changes are illustrated with a case history of Los Angeles smog.

Twenty years ago a manuscript on air pollution was apt to find a lonely burial place in *Engineering and Science*. Since that time we have come a long way. Today, the student of air pollution control has a wide choice of reading matter ranging from publications that are purely scientific and technical, to essays that discuss the social, economic, administrative, and legislative aspects of the subject.

Abatement strategy contains some of each one of these aspects. Medical input tells us how far to control; economic and technological considerations show how to do this; and administrative action, backed by law, puts the necessary controls into effect. As a corollary of this, the administrator had better be acquainted with some of the basics of technology as well as basics of legislation, and the engineer should be aware of some of the social and economic constraints. The medical expert might desire zero pollution but might ignore serious, economic consequences of such extreme policy.

The abatement strategy is the basis for a sound and successful attack on the pollution problem and starts by recognizing and assessing the problem. This is followed by developing guidelines for emission control: ambient air standards, emission standards, and applying appropriate control techniques. The control strategy aims at an air quality which will guarantee, at some time in the future, the preservation of health or the enjoyment of life. The methods to reach this goal must be technologically sound and economically and socially acceptable. This strategy is not static; it is a policy which has to be adjusted to technological developments and to changes in environment as well as attitudes of people. Our urban areas expand, and population density increases rapidly. The urban area as it spreads takes more and more room, and the demand for services increases as the population grows and becomes more affluent. Industries increase in number and in size, and many more automobiles fill the streets. However, some problems are solved and disappear; others are deferred to be solved later. A strategy has to be tested and reexamined to remain effective and useful.

Discussions on various aspects of control strategy have appeared in reports of symposia and committees; these are useful guides in the new field of planned pollution control (1, 2, 3, 4, 5). Here the experience of California in forming and executing its control strategy is used to illustrate the various facets of this process. In the past the pattern set in California has had a major impact on pollution control in other parts of the country, as well as on federal policies, in the developing of technical, administrative, and legal tools in combating photochemical smog.

History of Smog Control in Los Angeles

Los Angeles was the first city to recognize that it was afflicted with photochemical smog and for three decades has been in the forefront to fight the disease. In the early phase of the control effort, the problem was considered to be a local one which had to be solved by local agencies. This approach was justified since power plants, smelters, foundries, open dumps and incinerators, and the uncontrolled release of sulfur oxides from refineries were immediately criticized by an aroused public. To control a directly observable smoke plume, all that is needed to make the problem disappear are a few incensed citizens, a city council willing to listen, and the application of existing technology. This was not true with the problem which first developed in Los Angeles during the early forties. After smoke and oxides of sulfur were controlled, the pollution problem (known now as photochemical smog) persisted, and the only result which the citizens could see was that the black industrial cloud, which moved into the city from its southern industrial area, was replaced by a structureless, grayish, eye-irritating haze covering all Los Angeles and stretching far into the adjacent valleys of San Fernando and San Gabriel.

There had been some warning of the deteriorating of the atmosphere when in 1942 in the midst of World War II eye irritation first occurred. This was assumed to be coming from a synthetic rubber plant near the center of the city. The attacks disappeared near the end of the war when the plant was closed; however, this was followed by renewed and more widespread attacks of eye-irritating pollution, and many suspects were named. Refineries, chemical factories, open burning, and automobiles headed the list, but not one of these operations could explain the almost daily occurrence of the new pollution syndrome called smog. Smog, a contraction of smoke and fog, remained a household word even after the chemists established that the problem was of a totally different origin. Almost no coal was burned in the area, and sulfur dioxide had been controlled so the origin of the eye-irritating haze, accompanied by a peculiar bleaching-solution odor, was a disturbing mystery. More evidence that there was something seriously wrong with the air over Los Angeles soon appeared. Rubber manufacturers received complaints about rapid deteriorating of their products; plant scientists, John Middleton and Frits Went, detected widespread damage to soft-leaf vegetation not observed elsewhere (6).

From the analysis of air samples taken during smog periods, I concluded that the objectionable agents were formed in a photochemical reaction which led to the oxidizing of the ever-present hydrocarbons. The smog effect was readily reproduced in the laboratory by irradiating a mixture of hydrocarbons and oxides of nitrogen. A systematic analysis of the polluted air proved to be an essential part in the control strategy in Los Angeles, as in other areas plagued by photochemical smog.

Administrative Development in Control

Cities—*e.g.,* Pasadena which for a long time had been regarded as a haven for those who valued the quality of their mild climate and scenery —daily saw mountains disappearing in ugly haze. This, perhaps more than anything else, was responsible for a clean air movement by private citizens. This movement was joined by city officials who were aware that their city received the aerial garbage from adjacent Los Angeles. It soon became clear that one strong organization was needed in county government to cope more effectively with the widespread problem. This was a new phase in the control strategy: the transfer of authority to higher levels of government, a trend which has been consistently followed. The present eco-generation, pressing state and federal governments for action, hardly realizes that abdication of the self-determination rights of the individual community was almost revolutionary in the mid-1940's (*See* Reference 7). It was Harold W. Kennedy who was then the Los Angeles County's counsel who forged the legal tools for the County to resolve the problem of smog and the collision of governments and industry. The Constitution of the United States reserves the right of the States to use police power to protect the health, safety, and general welfare of the inhabitants. Based on the assigning of this right from the State to the county, the Board of Supervisors of Los Angeles County in 1945 created the office of Director of Air Pollution Control and adopted ordinances to prohibit the emission of excessive smoke.

Many cities in the County voluntarily joined in the battle against smog; however, much more was still needed. The air pollution problem superseded city boundaries, but the County could not enforce within city jurisdictions. There was a clear need for a single agency to be fully responsible for controlling air pollution throughout the entire county.

The State Air Pollution Control Act of 1946 (Stewart Bill) satisfied these conflicting demands by assigning jurisdiction for controlling air pollution to counties. This bill created in each county of the State an air pollution control district which could become a fully active control agency when so resoluted by the County Board of Supervisors when there was an air pollution problem that could not be solved by individual city and county ordinances. Immediately after this law was enacted, the Los Angeles County Air Pollution Control District (LACAPCD) was established. After some delay, showing the difficulty of concerted action of independent governmental agencies, Orange (1950), San Diego (1955), Riverside (1955), and San Bernardino (1956) counties formed their own districts.

Emission Inventories and Engineering Control

The first air pollution control officer of the LACAPCD was Louis McCabe who initiated stringent restrictions on emissions of sulfur dioxide and smoke from power plants, the petroleum industry, and the chemical and metallurgical industries. McCabe, who had foreseen the need to know more about the nature of the problem in Los Angeles, organized a research program which parallelled the control activities. This program confirmed that Los Angeles pollution was quite different from the smoke problems that were common to many industrial eastern cities. An extensive program was started to trace the origin of the primary reactants in the photochemical smog.

The Control District's testing crews began a detailed survey of the type and quantity of pollutants released from thousands of sources, large and small. This pollution inventory proved the large contributors of hydrocarbons to be the petroleum industry and the automobiles, and of oxides of nitrogen to be the automobiles and the fuel-burning power plants (8).

Gordon Larson who succeeded McCabe continued to reduce the emissions of refineries and other industries. He also undertook the control of emissions from open dump burning and backyard incinerators and made the initial contacts with the automobile industry to control motor vehicle exhaust emissions. However, in this drive to control the proved main sources of air pollution, Larson was opposed by the petroleum, chemical, and other industries, and primarily the individual citizen with his backyard incinerator.

Public Education and Enforcement

The pressures accumulating from so many sources were politically unacceptable. Larson's successor diagnosed the problem as lack of public support at this critical junction. Smith Griswold, the new director, saw the need for visible action and organized a strong enforcement and inspection team. The uniformed enforcement officers with sheriff's credentials, operating in clearly marked cars equipped with two-way radios, established much confidence in the control efforts of the District. The new control officer also recognized the great importance of a strong public education department which could inform the citizens of the nature of the problem and the steps taken to control it. Anyone who has tried to explain photochemical reactions to a public who demands action now will understand the need for education. Decision-making always clashes with special interest groups—e.g., industry, politicians, or private citizens. No control agency can escape the eroding criticism of these forces. One of the best counter-actions available to the enforcement agency is a strong public education and information program which prepares and informs the citizens that restrictions of their activities are unavoidable and in their best interests (9). These thoughts are still true today. Most local and state governmental agencies are severely hampered by budget restrictions in their educational mission.

The Role of State and Federal Governments

Although California law had assigned the major role controlling air pollution to the counties, practical experience taught that state government could not escape a major responsibility in some general aspects of this control. The forming of the Bureau of Air Sanitation in the State Department of Public Health foreshadowed the more prominent role which the State would play. This Bureau, directed by John Maga, was charged with studying of the causes of air pollution, with aiding county districts in abating air pollution, and with determining effects on health and other air pollution effects. In 1959 the Bureau set the first standards for ambient air quality and for motor vehicle emissions in California— a most significant conclusion to the first ten years of pioneering work.

In 1960, the State of California took over the responsibility of controlling motor vehicles by creating a Motor Vehicle Pollution Control Board. This was a logical move since moving sources, such as automobiles, do not respect local boundaries. This transfer of authority was also advantageous since the whole State could now pressure the automobile industry. The new Board was given the duties of curtailing the emissions from motor vehicles in an orderly way by adopting testing procedures followed by certifying of control devices to meet the emission standards which had been adopted in 1959.

The Board's efforts resulted in installing crankcase devices on California cars since the 1961 models and installing exhaust devices on California cars beginning with the 1966 models.

The Mulford–Carrell Act of 1967 dissolved the Motor Vehicle Pollution Control Board in California and created the Air Resources Board that was provided with broad powers and authority and with the ultimate responsibility of controlling air pollution in California. The California Air Resources Board divided the State into eleven air basins, areas with similar meteorological, topographical, and air pollution problems. Ambient-air quality standards were adopted which apply to all of these basins. The enforcing of these standards is still primarily a function of local go vernment, but emission control programs of local agencies have to be s ibmitted to and approved by the State Board.

Meanwhile the federal government had not been idle, and its activities betray a similar shift to a more centralized approach to environmental matters. The Federal Clean Air Act of 1963, its amendments in 1965, and the Air Quality Acts of 1967 and 1970 put the federal government prominently into the picture. The 1970 legislation requires the implementing of ambient air standards set by the federal Environmental Protection Agency. A strict and tight enforcement schedule is intended to solve the nation's pollution problems from stationary sources in three to five years.

The federal government also preempted the motor vehicle emission control. California alone, among all the states, was able to receive waivers of federal preemption of motor vehicle control. This was permitted because California was able to prove that it had a pressing need for vehicle emission standards more stringent than those adopted by the federal government for the remainder of the country.

Air Basin Strategy

The control of air pollution has gone through an evolutionary process. Slowly government has recognized that conserving the quality of air in one community cannot be separated from that of a neighboring one. The problems of Los Angeles City and those of the county have been immersed in the larger problems of the South Coast Air Basin. This basin philosophy was officially recognized by the action of State and Federal legislation, whereby future decisions on air pollution control will be increasingly based on their impact on the region rather than on a city or county.

Publications of the Air Resources Board give detailed information on: the extent of the various basins in California, the state ambient air standards, the emission standards for various model years of motor vehicles, the air pollutant emissions, emissions prevented, and air pollution levels (10).

Meteorological and geological features determine the extent of this region, the South Coast Air Basin. A series of mountain ranges, whose upper elevations reach over 10,000 feet above sea level, form a semicircular barrier surrounding the South Coast Air Basin. A gentle sea breeze moves pollutants inland during the daytime, and an even weaker land wind reverses the smog cloud during the evening and night. Temperature inversions during most of the year aggravate the normal condition of poor ventilation in the region. In this big fumigation room of 8,700 square miles, 10 million people with their industry and their 5 million cars daily disperse 11,000 tons of carbon monoxide, 3,200 tons of hydrocarbons, and 1,500 tons of oxides of nitrogen (1970 data).

Criteria and Ambient Air Standards

As a basis for all efforts of air pollution control, the California Air Resources Board, as well as the Environmental Protection Agency of the federal government, has adopted ambient air quality standards based on health and aesthetic considerations. These standards have been derived from observing living organisms in the laboratory as well as in the community. These studies were compiled in the criteria documents of state and federal control agencies (11).

How clean we want the air is a most important, but highly subjective judgment. The ultimate in control is no pollution at all. However, the cost of this ideal would certainly result in the downgrading of other desirable community services: schools, transportation, safety, etc. Pollution control must take its place among the desirable tax-consuming, community activities and is always subjected to a relative importance judgment (12, 13). The subjective nature of these important decisions are clear from the wording accompanying the establishment of the federal oxidant standards. This index of photochemical eye irritation is set at 0.08 ppm hourly maximum which should not be surpassed for more than one day per year.

Source Emission Standards

Once we set acceptable ambient air standards, the control agencies must convert these into a message that can shape a strategy so that the ambient air standards will be met. Various levels of government have been active in setting these emission standards. Some agencies—e.g., the County of Los Angeles—have done this for stationary sources by adopting a set of administrative rules (14). The State of California has set standards for automobile emissions through its Air Resources Board as well as through direct legislative action. The Federal Air Quality Act of 1970 contains the latest and most stringent set of these standards for automobiles (15).

The different governmental approaches to setting emission standards supplies interesting debates. At first the standards were set by governmental agencies; recently legislatures have been writing the motor vehicle emission standards into law. Standards set by legislatures are less vulnerable than those set by an agency; however, they are less flexible when adjustments have to be made because of better knowledge of criteria or of the increased potential of our technology or special conditions of a local nature. In the early phase of controlling automobile exhaust, the state legislative program up to the 1974 model cars had a revolutionary effect on the activities of the automobile industry. The same applies to the federal program which has set automobile standards for the 1975 and 1976 model cars. Many technically-oriented experts question the attainability of these standards. Any readjustment or correction can only be accomplished in Congress by a most difficult and politically unattractive exposure. The process by which governments have arrived at emission standards has taken many forms and often leaves much to be desired, largely because of the paucity of available data. Emission standards interfere directly with business-as-usual whether it is the operation of an industry or an automobile, and heated discussions and controversies are normal in this field (16).

In our urban areas, we recognize two main types of pollution: one coming from single sources—e.g., power plants and foundries, the other coming from multiple sources—e.g., the automobile. For the single, isolated sources, the ambient air standards are translated without too much difficulty into source emission standards. We have reasonably reliable mathematical formulae and empirical data to predict ground concentrations at some distance from point sources such as power plant stacks, and these calculations on the average indicate the true pollution potential. However, in a more densely populated area this average might not be accepted when abnormal wind patterns cause a downdraft of the pollutants, upsetting those who live close to the plant. The principle usually followed by the control authorities is to set emission standards which require using the best available control technique. However, even a simple-appearing problem—*e.g.*, the impact of emissions of sulfur dioxide on the surrounding area—turned out to be more complicated than we had assumed. Sulfur dioxide does not remain unchanged after it has left the stack; it is slowly converted into visibility-obstructing sulfates, and this process is hastened by the photochemical pollution complex. But this is not all; it is now well established that the effect of the sulfur oxides is enhanced by the presence of particulate matter. This is held responsible for some of the often-cited air pollution disasters such as occurred in London and in Donora.

Although the chemistry of these pollution episodes is different from that caused by photochemical pollution, photochemical pollution also produces particulates, and it is reasonable to expect a similar potentiating or synergistic effect of various pollutants.

The problem in this case is even more complicated because we are considering the interaction of oxides of nitrogen, sunlight, and organic materials, mostly of gasoline origin, consisting of many different components—*e.g.*, saturated hydrocarbons, olefins, aromatics, and their oxygen derivatives of different chain lengths and builds. The concentration and relative proportions of the emissions depend highly on time and place of release—*e.g.*, the corner of a busy intersection, the freeway on commute hours, superimposed by a constant stream of oxides of nitrogen from stationary sources. Among the many unpleasant and complicating features that photochemical smog brings, there is at least one aspect which facilitates the estimation of needed control: the reactions leading to the smog symptoms caused by the forming of secondary products need considerable time, and as a consequence the pollutants are reasonably well mixed before the peak oxidant values and irritating effects are reached.

Air Basin Capacity and Model Calculations

In calculating emission standards, we must consider not only the effects close to the source but also those at far greater distances caused by pollutants gradually filling the basin. This becomes especially important when meteorological and geographical conditions limit the dispersing of the pollutants and the capacity of an area to accommodate pollutants without exceeding the ambient air standards is greatly reduced. The capacity of the South Coast Basin in which Los Angeles is situated varies with the height of the inversion. For example, when the temperature inversion is at 500 feet, about half of the basin is below the inversion (assuming that this area slopes lineally from sea level to 500 ft). The total volume of the region below the inversion layer is about 150–300 cubic miles. When the inversion height doubles, the volume of the basin increases too, and more pollutants can be tolerated.

An empirical way of calculating this critical pollutant capacity of the basin consists of comparing the ambient air concentration of a stable pollutant—e.g., carbon monoxide—during heavy smog with the amount released in the basin. For example, the release of 11,000 tons of carbon monoxide needs to be dispersed in 200 cubic miles of air to yield an average ambient air concentration of 10 ppm which is about the mean concentration measured in the basin. During severe pollution periods the dispersal volume of the basin is much smaller, and hourly concentrations of 50 ppm have been measured. From this critical pollution volume and the frequency of low inversion, the required control is estimated. The estimate takes into account the stability of the various pollutants and the time needed for the control. Even though inaccurate, these calculations do point to the need for 80–90% control—a result which agrees with the roll-back method most often used in calculating the required degree of control.

Scientifically, we would like to consider the basin as a box of known dimensions in which various pollutants are released to give predictable concentrations as was done in the simple calculation above. However, the problem is much more complicated. The basin is not a closed system; wind patterns and wind strength vary; inversion conditions change continuously and are not even the same for the whole basin. The uneven distribution of emissions and subsequent atmospheric reactions add numerous complications. Much progress will be made in the science of modelling. Now, it is still an art and only applies to pollution cases less complicated than we find in the Los Angeles Basin. Nevertheless, a blend of model experiments, empirical input of tracer experiments, and monitoring data can greatly add to estimating the capacity of a basin (*17, 18, 19*).

Roll Back Method

Now we have to rely largely on an empirical approach which assumes that the emissions are proportional to the measured ambient air values. This is true for stable pollutants such as carbon monoxide, but it is not true for secondary pollutants—*e.g.*, nitrogen dioxide or oxidants. Nevertheless, as a first approximation and by using restraint in the interpretation, the so-called roll back method, based on this proportionality of emissions and measured ambient air standard, is most often used.

If the present ambient air level is A and the legal level is B, the degree of control needed is $(A - B)/A$. When background concentrations are significant as in oxidant measurements, these should be noted, and the roll back formula takes the form of:

$$\frac{(A-C) - (B-C)}{(A-C)}$$

The maximum nitrogen dioxide concentration measured in the South Coast air basin in 1970 was 0.83 ppm, hourly average. Applying these formulae, we find that we need to reduce concentrations by 70% to stay within the legal limit of air quality. When we consider that it may take as much as ten years to accomplish this, we must aim at more stringent control and apply a growth factor. The planned control of oxides of nitrogen therefore should not be less than 80%.

These estimates are based on reaching the health standard of 0.25 ppm; however, the nitrogen oxides also are dominant in the forming of photochemically-created toxic materials. Here, we must consider the simultaneous presence of two types of components, the oxides of nitrogen and the organic material. In this complicated, multi-dimensional system case, we should consult the empirical relations established among hydrocarbons, oxides of nitrogen, and the resulting symptom, oxidant or eye irritation. Comparing these data on smog and non-smog days will determine the measure of control needed. In Figure 1, the data on the concentration of the oxides of nitrogen and hydrocarbons at 7:30–8:30 AM on a smog-free day when the oxidant was less than 0.1 ppm are compared with a severe-smog day when the oxidant exceeded 0.4 ppm.

The position of the areas of high and low oxidant agrees well with the results obtained in human fumigation experiments, as indicated by the iso-irritaton lines (*20, 21*). The oxidant values, less than 0.1 ppm, fall in the no-smog area; those above 0.4 ppm fall mostly in the medium and severe area. It is gratifying that the emission data are lined up in a ratio predicted by basin pollution inventory measurements. These three independent types of observations agree satisfactorily, giving confidence in using these results for the calculations of the roll back calculations. These data show that presently we should require a reduction of at least 75% of both smog components. Figure 1 shows that a limited trade-off is possible whereby the lesser control of the oxides of nitrogen is made up by a more stringent hydrocarbon control. The limit of this trade-off is determined by the 0.25 ppm NO_2 health standard.

To keep up with the increase in human activities, a growth factor must be applied as in the examples mentioned earlier. This correction factor differs for various classes of emitters. For automobile emissions, a ten-year growth of 2% per year requires an adjustment by a factor of 1.2; for power plants, the 100% growth in the same period requires a correction factor of 1.5.

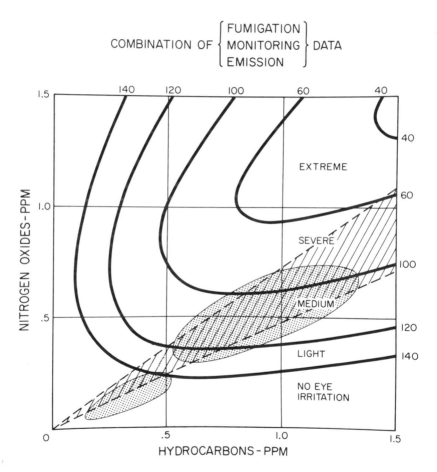

Figure 1.   Calculation of roll back in a photochemical system of two components

The small ellipses contain air quality data during days with no eye irritation with oxidant values of less than 0.1 ppm. The large ellipse contains values during severe smog with oxidant values higher than 0.4 ppm. The hydrocarbon concentrations are erpressed in ppm hexane.

Automobile Emission Standards

These calculations have not taken into account the realities of technological problems in mass producing vehicles. For example, the prototype automobile developed in the laboratory is not the same car which the consumer is going to drive. The prototype is a laboratory product constructed carefully and adjusted to obtain minimum emissions. On the way to a salable product, adjustments have to be made to conform to the technology of mass production. To reach the required standard at the end of the assembly line, the prototype must have considerably lower levels of emission than the finished products. Decisions have to be made on the questions of durability, maintenance, and allowable deviations from the average. To the factors on production slip, quality control, and durability, we add an insurance factor. This is the factor so well known to chemists and usually expressed as the law of maximum unhappiness which predicts that if something can go wrong it usually does.

In these calculations the time element should be important. Just as the cost increases drastically by increasing the degree of control beyond a certain point, the same applies when the manufacturer is pushed beyond a reasonable time schedule to meet the standards. Once the car has left the assembly line, the control agency has to make sure that the control systems function properly for a reasonable time. Inspection is necessary for the control effort. In an area of high photochemical pollution potential, there is a need for retro-fitting used cars, to shorten the time required for meeting the legal ambient air standards. Each one of these requirements has a pronounced impact on society by raising the purchase price of cars, the cost of maintenance, and fuel use. The developmental engineering cost for meeting the progressively more strict standards has taxed some of the smaller companies to the breaking point, and an undesirable by-product of too stringent regulations is the removal of smaller companies from the competition whereas the larger companies with their extensive research potential are able to meet the standards. These considerations and many others of social and economic importance have to be weighed by governmental agencies when developing emission standards and are essential in the overall control strategy (22, 23).

In a subject where goals cannot be precisely defined, gray areas exist which are subject to personal judgment. It is not too surprising, therefore, that there are those who demand zero pollution while others are content with lesser purity. As Paul Kotin once said, "It all depends on how evangelistic you are."

Unless a good deal of common sense is used in establishing the right blend of stringency, too weak or unnecessary repressive measures may result. The setting of these standards is a most important part of the control strategy, and misjudgment can be costly, running into millions or billions of dollars. Calculations are long overdue on the optimum strategy to reach these low emission limits at the least expenditure of resources.

Long Range Strategy

This need is especially pressing when we look beyond a short range goal of a few years, when we want to be sure that ten or twenty years hence the air will be of acceptable quality. At the present rate of growth, the population in California will have grown from 19,703,000 in 1970 to 23,249,000 in 1980, the South Coast Basin, from 9,717,000 to 11,300,600 (24). During that time we will have doubled the use of electric power. Parallel with this increase goes the establishment of industry and all the polluting activities associated with the normal operations in an urban community. Figure 2 shows how the oxides of nitrogen start increasing again after a decrease from a maximum in 1970 to a minimum in 1985 (25). In 1985 the miscellaneous sources—i.e., house heating and numerous other small sources—will already comprise more than the total allotted maximum emissions for the South Coast Basin. This does not even leave room for the controlled emissions of cars and power plants.

Acceptable limits of water, land, and air pollution could be achieved if costs were not considered. However, economics is a part of life and in many areas pollution has advanced to the extent that tens of billions of dollars are needed to catch up with the problem, and as the population of the United States increases and more demands are placed on our resources, more sophisticated techniques will be required to reach acceptable air quality levels. Whatever the outcome of these computations are, any implementation plan will meet with almost insurmountable obstacles. This is true for meeting the California standards plan but even more so for the federal ones.

A national resources study carried out under the Academy Resources Council sponsorship in 1961–62 summarized and recommended that efficient planning and wise management of our natural resources require a much broader and more fundamental understanding of the dynamic relationships which couple all elements of a resource system from first discovery to eventual use in an international economic system. The process of decision-making as it relates to resource use deserves a searching examination, including the roles of public attitudes, laws, and political and social organizations. The factors that stimulate and guide technological development, including economic motivation and influences of social setting, should also be determined. The means of estimating

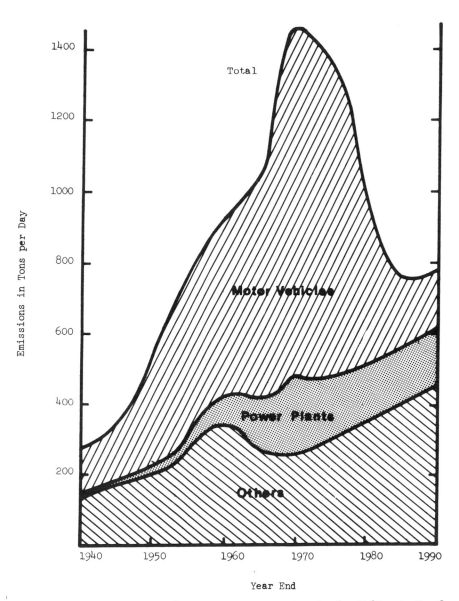

Figure 2.   Estimates of oxides of nitrogen emissions in the California South Coast Basin

economic efficiency of alternative methods of resource use are far from satisfactory; sounder techniques should be devised. A comprehensive study under private auspices, but with full support from government, should be undertaken, centered on a selected specific problem (3).

This is not the first time that a warning has been sounded. Several years ago P. A. Leighton of Stanford University predicted that eye irritation in 1980 would be about the same as it was in 1960 (26). Leighton wrote, "In a sense our pollution may be likened to a weed. Controls may clip back the weed, but will not keep it from growing again. To kill the weed we must get at the root, and the root of the whole problem of general pollution is combustion. . . . The proper approach to a lasting solution of these problems, the only way to kill the weed, is to attack, not the products of combustion but combustion itself, to reduce by every possible means the burning of fuels in favor of non-polluting sources of heat and power" (26).

All our available technology must be mobilized to meet this challenge. Research organizations in government, universities, and industries all need to participate in finding ways to improve the technology of our pollution control. As long as we do not have such improved control technology, our only choice is to limit expansion of all polluting activities in the basin.

The plans have to be made now as to how we want our urban areas to look and what kind of air we want to breathe a few decades ahead. This plan of action must include conserving green areas, developing a comprehensive, non-polluting transportation system, and managing our industrial complex, with an overriding regard for its effect on the environment. We will be forced into making drastic and revolutionary decisions in community planning which involves judicious locations or relocation of residences and industries to areas having better ventilation. It might have to face the gigantic task of converting the area to a totally electric economy, eliminating all fuel burning in the basin, and using only electric power which should be generated outside the basin for propulsion and for space heating.

In the past much progress has been made in pollution control, and much has been learned; attitudes have changed; ecology has become a household word; and almost everybody agrees that something should be done about pollution. The noise level is rather high at present, but what really counts is how much profit are we willing to forego, how much change in our accepted way of living are we willing to accept in exchange for breathing clean air. This means a radical break with ingrained social and economic behavior and means a struggle against almost insurmountable odds, which make the controversies of the past look like a tea party. Now is the time to do the planning. Every day we wait new decisions will have been made which make a sensible solution to the environmental problem more difficult. Let us get on with the job; we have no time to lose.

Literature Cited

1. Atkinson, A., Gaines, R. S., Eds., "Symposium on the Development of Air Quality Standards," Merrill, Columbus, 1970.
2. Barnett, H. T., Chandler, M., "The Economics of Natural Resource Availability," Resources for the Future, Johns Hopkins, Baltimore, 1969.
3. "Waste, Management and Control," Publication 1400; National Academy of Sciences National Research Council, Washington, D. C., 1966.
4. Wolozin, H., Ed., "The Economics of Air Pollution," Norton, New York, 1966.
5. Stern, A. C., Ed., "Air Pollution," Academic, New York, 1968.
6. "Recognition of Air Pollution Injury to Vegetation, A Pictorial Atlas," J. S. Jacobson and A. Clyde Hill, Eds., Air Pollution Control Association, Pittsburgh (1970).
7. Kennedy, H. W., "The History, Legal and Administrative Aspects of Air Pollution Control in the County of Los Angeles" (May 9, 1954).
8. Second Annual Report of the Los Angeles County Air Pollution Control District, 1952.
9. Maga, J. A., "Air Pollution," 2nd ed., Vol. 3, pp. 797–811, A. Stern, Ed., Academic, New York, 1968.
10. Publications by the Air Resources Board and the Resources Agency of the State of California: California Air Basins (May 1969); Ambient Air Quality Standards (January 1970); Control of Vehicles Emissions after 1974. Report to the California Air Resources Board by the Technical Advisory Commission (Nov. 19, 1969); Emission Inventories (November 1969); Air Pollution Control in California (January 1970).
11. "Criteria Documents," National Air Pollution Control Administration Publications: Air Quality Criteria for Carbon Monoxide (March 1970); Air Quality Criteria for Nitrogen Oxides (January 1971); Air Quality Criteria for Photochemical Oxidants (March 1970).
12. Kohn, R. E., "Symposium on the Development of Air Quality Standards," pp. 103–125, Chapt. V, A. Atkinson and R. S. Gaines, Eds., Merrill, Columbus, 1971.
13. Stern, A. C., "Air Pollution," Vol. 3, A. Stern, Ed.
14. Stern, A. C., "National Emission Standards for Stationary Sources," *J. Air Pollution Control Assoc.* (1970) **20**, 524.
15. Barth, D. S., "Federal Motor Vehicle Emission Goals for CO, HC, and NO_x, Based on Desired Air Quality Levels," *J. Air Pollution Control Assoc.* (1970) **20**, 519–523.
16. Heuss, J. M., Nebel, G. J., Collucci, J. M., "National Air Quality Standards for Automotive Pollutants—A Critical Review," *J. Air Pollution Control Assoc.* (1971) **21** (9), 535.
17. Wanta, R., "Air Pollution," pp. 187–226, 2nd ed., Vol. VI, A. Stern, Ed., Academic, New York,
18. Reignam, H., "Atmospheric Environment," pp. 233–247, Pergamon, New York, 1970.
19. Seinfeld, J. H., "Air Pollution," pp. 169–206, Chapt. VII, A. Stern, Ed., Academic, New York, 1970.
20. Haagen-Smit, A. J., "Urgent Problems in Air Conservation," University of Wisconsin Pilot Project in Environmental Sciences (Jan. 10, 1968).
21. Haagen-Smit, A. J., "Science, Scientists and Society," W. Beranek, Ed., Bogden and Quigley, in press.
22. Ridker, R. C., "The Economics of Air Pollution," pp. 87–101, H. Wolozin, Ed., Norton, New York, 1966.
23. Lamale, Helen H., *Ibid.*, pp. 115–126.

24. "Population Estimates for California Counties," Department of Finance, State of California (Sept. 15, 1971).
25. "Air Pollution Control in California—1970 Annual Report," The Resources Agency, State of California, Sacramento (January 1971).
26. Leighton, P. A., "Man in California, 1980's, Man and Air in California," pp. 84–116 (Jan. 27, 1964).

RECEIVED June 2, 1971.

Ozone Reactions in the Atmosphere

A symposium co-sponsored by
the Division of Industrial
and Engineering Chemistry and
the Division of Water, Air, and
Waste Chemistry at the 161st
Meeting of the American
Chemical Society, Los Angeles,
Calif., April 1, 1971.

Lyman A. Ripperton
Symposium Chairman

PREFACE

The papers in this symposium deal with reactions of ozone relative to the lower atmosphere, or troposphere, rather than the stratosphere. In the troposphere ozone is generated and destroyed. Synthesis of ozone is most spectacular in polluted urban air where surface concentrations have been recorded which are an order of magnitude greater than those found in the air of remote sites.

The processes in which ozone is a reactant, instead of a product, are important in removing ozone from the atmosphere and determining the behavior of many other atmospheric trace constituents. Ozone is one of the key substances in the chemistry of air pollution and in the chemistry of the natural atmosphere as well.

Until the discovery that it existed in urban air in parts per million and even tens of parts per million concentration, tropospheric ozone was largely in the domain of the meteorologist and the atmospheric physicist. The discovery of high urban concentrations of ozone stimulated a great deal of chemical research. The reason for the intense interest was that molecule for molecule, ozone is one of the most toxic inorganic substances known. Observed, ambient concentrations were reaching levels which could be injurious to man and which could be demonstrated to be destructive of some types of material. Although ozone is perhaps the most intensively studied trace constituent of the atmosphere, it still manages to surprise the researcher.

The papers presented here represent current research into the problems of the chemical behavior of atmospheric ozone and indicate the scope of the reaction involved.

Lyman A. Ripperton

Chapel Hill, N. C.
September, 1972

The Reaction of Ozone with Ammonia

KENNETH J. OLSZYNA and JULIAN HEICKLEN

Department of Chemistry and Center for Air Environment Studies,
The Pennsylvania State University, University Park, Pa. 16802

Mixtures of O_3 and excess NH_3 react at $\sim 30°C$ to produce O_2, H_2O, N_2O, N_2, and solid NH_4NO_3. The amounts of the gas-phase products, relative to the ozone consumed, are 1.05, 0.31, 0.032, and 0.031, respectively. Neither H_2 nor NH_4NO_2 was produced. For $[NH_3]/[O_3]_0$ ratios < 50, the disappearance rate of O_3 was first order in $[O_3]$ and increased slowly with increasing $[NH_3]/[O_3]_0$ to an upper limiting value of 0.21 min^{-1}, where $[O_3]_0$ is the initial pressure of O_3. As the reaction proceeded and the $[NH_3]/[O_3]$ ratio passed 120 (or if $[NH_3]/[O_3]_0 > 120$), the rate shifted to three-halves order in $[O_3]$ and was proportional to $[NH_3]^{-1/2}$. The reaction is interpreted as a chain mechanism with the heterogeneous decay of O_3 as the initiating step. Nitrogenous products come from oxidation of HNO with O_3, followed by reaction with NH_3.

The room-temperature photooxidation of NH_3 has been studied by several workers, and the results have recently been reviewed (1). The products of the reaction are H_2, N_2, H_2O, and NH_4NO_3. Bacon and Duncan (2) reported that the overall reaction is approximately represented by the expression

$$8NH_3 + 7O_2 \rightarrow 2N_2 + 2NH_4NO_3 + 8H_2O$$

Apparently the only study of the ozonation of NH_3 was done by Strecker and Thienemann, and they found that the products are H_2O, O_2, NH_4NO_3, and NH_4NO_2 (3). They reported the overall stoichiometries to be

$$2NH_3 + 4O_3 \rightarrow 4O_2 + H_2O + NH_4NO_3$$
$$2NH_3 + 3O_3 \rightarrow 3O_2 + H_2O + NH_4NO_2$$

Because of the lack of data on the O_3–NH_3 reaction and because both gases are significant impurities in urban atmospheres, we reinvestigated this system. We analyzed the products and made kinetic measurements. The results are reported below.

Experimental

Most of the gases used were from the Matheson Co.; these included extra dry grade O_2, C. P. grade NH_3, and prepurified H_2. The ultra pure grade helium was used after passing through a filter containing glass wool, Drierite, and Ascarite. The NH_3 was distilled from $-100°$ to $-196°C$; the O_2 and H_2 were used after passing through traps at $-196°C$. The only detectable impurities were 0.05% N_2 in the O_2, < 0.01% air in the H_2, and none (< 0.01%) in the NH_3.

Ozone was prepared from a tesla coil discharge through O_2. The ozone was triply distilled at $-186°C$ and collected at $-196°C$ with continuous degassing.

The ozone, hydrogen, and oxygen pressures were measured by an H_2SO_4 manometer; the O_3 pressure was checked by optical absorption at 2537 A. The NH_3 and He pressures were measured by an Alphatron Vacuum Gauge, Model 820.

The gases were introduced into the cell, and the reaction was monitored continually by ultraviolet absorption spectroscopy using low intensities so that photochemical reaction was not induced by the monitoring lamp, which was a Philips 93109E low-pressure mercury resonance lamp. The radiation passed through a Corning 7-54 filter to remove radiation below 2200 A and above 4200 A and a cell filled with chlorine to remove radiation above 2800 A before passing through the reaction vessel to a RCA 9-35 phototube. Some runs were monitored with radiation between 3000–4200A where neither O_3 nor NH_3 absorb; this spectral region was isolated by a Corning 7-54 filter and a glass plate.

All gases were handled in a grease-free, high vacuum line using Teflon stopcocks with Viton "O" rings. The reaction vessel was a quartz cell 5 cm long and 5 cm in diameter. To test the effect of surface area and volume, runs were done in the vessel unpacked, as well as packed with 1000 solid spherical glass beads (Kimax, 3 ± 0.5 mm diameter). Adding the beads increased the surface area from 139 to 412 sq cm and reduced the volume from 102.3 to 88.1 cc. The beads were conditioned by ozone until the heterogeneous decay of pure ozone was the same as in the unpacked cell.

In most of the experiments after the reaction was complete, the products were collected and analyzed by gas chromatography. Aliquot portions of the noncondensable gases were analyzed on an 8 ft long 5A molecular sieve column at 0°C with an He flow rate of 94 cc/min. The condensable gases were distilled from $-100°$ to $-196°C$ and then from $-130°$ to $-196°C$. The residue from the $-100°C$ distillation and the distillate from the $-130°C$ distillation were separately passed through a Porapak T column 8.5 ft long operated at 79–82°C and a He flow rate of 140 cc/min. For both fractions a Gow Mac Model 40-012 voltage regulator with a thermistor detector was used with a 1-mv recorder.

The solid product of the reaction was identified by infrared spectroscopy (IR). The reaction was carried out in a borosilicate glass cell 6 cm long and 2.1 cm in diameter with NaCl windows but using higher

O_3 pressures and $[NH_3]/[O_3]_0$ ratios from 5–12. When NH_3 was added to the borosilicate glass cell containing ozone, a white aerosol, which deposited on the cell walls and windows, appeared almost immediately and remained even after the cell was evacuated. However for $[O_3]_0 \geqslant$ 10 Torr, an explosion resulted when NH_3 was added.

Results

When NH_3 is added to O_3 in the reaction vessel, the O_3 is consumed, and a white solid aerosol is produced which settles on the cell walls. The O_3 consumption is monitored by its intense absorption at 2537 A. For a freshly cleaned cell some decay of the O_3 occurs even in the absence of NH_3. However the reaction with NH_3 is much more rapid but is not reproducible. As the solid deposit accumulates from several runs, the background reaction becomes negligible, and the reaction with NH_3 becomes slower and reproducible. All reported results here are for the cell (packed and unpacked) conditioned this way.

The light monitoring the O_3 was too weak to cause photodissociation to occur significantly. This was checked by doing duplicate runs with continuous or intermittent radiation; decay curves were the same. Ozone decay curves were obtained for initial ozone pressures, $[O_3]_0$, of 0.14–3.02 Torr, initial NH_3 pressures of 0.36–292 Torr, and $[NH_3]/[O_3]_0$ ratios of 0.9–649. The decay curves were tested for the reaction order in $[O_3]$. The order depended only on the ratio $[NH_3]/[O_3]$ and varied between 1–1.5. For three runs at $[NH_3]/[O_3]_0$ of 10.5, 50, and 625 the first and three-halves order plots are shown in Figures 1 and 2, respectively. The results in Figure 1 show that at low $[NH_3]/[O_3]$ ratios ($[NH_3]/[O_3]_0 = 10.5$), a good first-order decay plot is obtained for at least 81% decomposition. In Figure 1 the ordinate scale for curve A is on the right for clarity; the numbers by the data points indicate percent O_3 decomposed. At an $[NH_3]/[O_3]_0$ ratio of 50, the curve starts to give a first-order plot but falls off from linearity above 60% decomposition where $[NH_3]/[O_3]$ becomes large (> 120). For an $[NH_3]/[O_3]_0$ ratio of 625, the first-order plot has no linear region. The same three runs are shown in Figure 2 on a three-halves order plot (ordinate scale on right for clarity; numbers by data points indicate % O_3 decomposed). The run at $[NH_3]/[O_3]_0 = 10.5$, which showed good first-order behavior, does not give a linear 3/2-order plot. The run at intermediate $[NH_3]/[O_3]_0$ does not fit 3/2-order initially, but after about 70% decomposition when $[NH_3]/[O_3] \sim 150$, the 3/2-order plot becomes linear. The run with $[NH_3]/[O_3]_0 = 625$ fits the three-half order law well to at least 88% decomposition. All of our runs showed the same behavior which can be summarized as follows.

$$-d[O_3]/dt = k[O_3] \qquad\qquad [NH_3]/[O_3] < 50 \qquad\qquad \text{(a)}$$

$$-d[O_3]/dt = k'[O_3]^{3/2} \qquad\qquad [NH_3]/[O_3] > 120 \qquad\qquad \text{(b)}$$

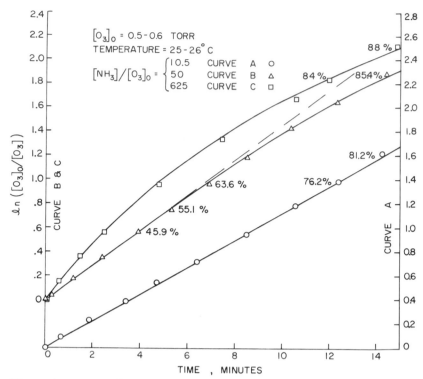

Figure 1.   First-order plots of the ozone decay in the NH_3–O_3 reaction at room temperature in the unpacked reaction vessel

It should be realized that the coefficients k and k' are functions of $[NH_3]$ which remained nearly constant during any run. These coefficients are listed in Table I for the unpacked reaction vessel and in Table II for the reaction vessel packed with 1000 glass beads.

After the optical density no longer changed, the cell was evacuated, and the gases were analyzed. The optical density did not change upon evacuating the cell but was slightly higher (0.01–0.04) than before the runs because of the absorption of the solid which did not pump away. Periodically the cell was cleaned by introducing about an atmosphere of NH_3 which removed the solid and reduced the absorbance. The decay curves were corrected for this residual absorbance in computing rate coefficients.

The optical attenuation resulting from the solid was more clearly shown by monitoring some runs with radiation between 3000–4200 A (principally 3660 A), where the O_3 absorption is unimportant. Some results are shown in Figure 3. Initially the optical attenuation increases to a maximum as the aerosol is formed and then drops as the aerosol deposits on the wall. The results are similar in the unpacked and packed cell. The presence of excess He (not shown) also hardly affects the results. The increase in optical attenuation is most marked when $[NH_3]$ and $[O_3]_0$ are large; however, correcting the decay plots is unimportant because the O_3 absorbance is also large. If $[NH_3]$ and $[O_3]_0$ are low, the aerosol interference is negligible (presumably because of more rapid settling on the walls). Optical attenuation because of the aerosol only interferes with the O_3 decay curves for high $[NH_3]/[O_3]_0$.

The solid product was identified as NH_4NO_3 by IR by performing some runs in cells with NaCl windows and pumping away the residual gases. The infrared spectrum showed the strong, broad absorption in the 3.03–3.30μ (3300–3030 cm^{-1}) region resulting from NH stretching vibrations of the NH_4^+ ion and the strong, broad NH_4^+ bending band near 7.00μ (1429 cm^{-1}). The weak, sharp absorption near 11.50–12.00μ (870–833 cm^{-1}), which is the stretching of the single bond of the NO linkage in the NO_3^- ion, is also present (4). The two strong N=O stretching bands in the 5.95–6.06μ (1680–1650 cm^{-1}) and 6.16–6.21μ (1625–1610

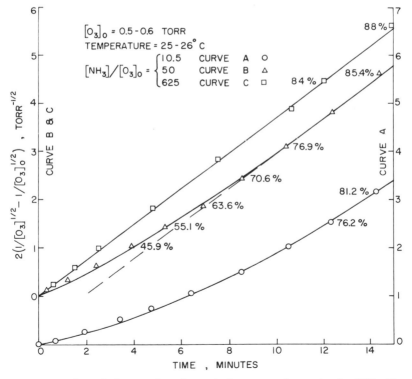

Figure 2.   Three-halves order plots of the ozone decay in the NH_3–O_3 reaction at room temperature in the unpacked reaction vessel

Table I. Reaction of NH_3 with

$[NH_3]/[O_3]_0$	$[O_3]_0$, Torr	$[NH_3]$, Torr	Temp, °C	k, min^{-1}	k', $Torr^{-1/2}\ min^{-1}$
0.9	0.39	0.36	32.3	0.084	—
1.2	1.52	1.8	28.1	0.106	—
1.8	0.20	0.36	31.5	0.106	—
1.85	0.62	1.15	26.4	0.083	—
2.8	0.39	1.1	31	0.105	—
3.4	0.81	2.8	27.9	0.125	—
3.9	0.79	3.1	30	0.125	—
4.3	0.14	0.61	31	0.126	—
4.4	1.44	6.6	26.8	0.141	—
5.0	1.14	5.7	—	0.141	—
6.1	0.78	4.8	28.2	0.143	—
6.5	1.23	7.9	—	0.149	—
7.0	1.03	7.2	—	0.160	—
8.3	0.78	6.5	31.5	0.178	—
13.1	3.02	39.5	30	0.184	—
13.6	0.99	13.4	32.0	0.185	—
14.6	2.06	30.1	30	0.197	—
15.2	1.80	28.3	32	0.203	—
16.6	1.49	25.9	26.7	—	—
18.1	0.75	13.6	31.9	0.195	—
21	1.29	27	31.9	0.202	—
22	0.40	8.7	33.8	0.183	—
22	1.66	38	32.4	0.210	—
22.7	2.08	47.1	24.3	—	—
23.1[a]	0.99	22.8	—	—	—
23.6	0.34	8.1	—	—	—
30	0.83	25.6	31	0.193	—
30	1.63	48	—	0.214	—
35	1.46	52.1	27.0	0.188	—
45.6	0.79	36.0	31.0	0.202	—
44.1	1.43	63.0	30.6	~0.23	0.37
50	0.48	24.1	24.8	—	—
51.6	0.45	23.3	29.9	~0.22	0.77
75	1.37	102	33.2	~0.30	0.26
80	0.47	37.5	30.8	~0.23	0.58
80	1.58	128	28.5	~0.24	0.24
88	0.86	76	29.2	~0.19	0.29
93	0.89	83	31	~0.25	0.34
118	0.44	53	26.8	—	0.33
119	0.38	45	26.2	—	0.34
130	0.66	86	30.5	—	0.30
495	0.43	213	27.0	—	0.39
649	0.45	292	26.3	—	~0.45

[a] O_3 decay not monitored; monitoring lamp off.
[b] Excludes runs with $[NH_3]/[O_3]_0 > 100$

O_3 in the Unpacked Cell

$[O_2]/[O_3]_0$	$[N_2]/[O_3]_0$	$[H_2O]/[O_3]_0$	$[N_2O]/[O_3]_0$
0.94	0.031	—	—
1.11	0.020	0.30	0.026
0.96	0.038	—	—
1.13	0.037	0.19	0.030
0.99	0.040	—	—
1.12	0.023	0.29	0.039
—	—	—	—
—	—	—	—
1.09	0.019	0.28	0.030
1.03	0.019	—	—
1.11	0.024	0.32	0.041
1.05	0.020	—	—
1.05	0.021	—	—
—	—	—	—
1.06	0.020	—	—
1.03	0.024	—	—
—	—	—	—
1.11	0.027	0.32	0.029
—	—	—	—
1.04	0.028	—	—
1.06	0.033	—	—
1.03	0.031	—	—
1.05	0.034	0.33	0.023
1.11	0.033	0.37	0.033
1.08	0.044	0.29	0.041
—	—	—	—
1.00	0.036	—	—
1.16	0.041	0.36	0.026
—	—	—	—
—	—	—	—
1.07	0.049	0.31	0.041
—	—	—	—
1.00	0.043	—	—
—	—	—	—
1.09	0.020	0.31	0.027
1.00	0.040	—	—
—	—	—	—
1.05	0.058	0.35	0.044
1.14	0.062	—	0.052
1.00	—	—	—
1.01	0.077	0.36	0.047
1.11	0.086	—	0.055
Ave. = 1.05	0.031	0.31	0.032
±0.05	±0.008 [b]	±0.03	±0.006 [b]

cm^{-1}) regions, which indicate nitrites, are absent. The nitrite absorption bands are among the strongest observed in IR spectra (4). Thus our results suggest the absence of NH_4NO_2, contrary to the findings of Strecker and Thienemann (3). The infrared spectrum obtained here agrees well with the Sadtler standard spectrum of NH_4NO_3 except for the strong, sharp peak at $\sim$3.3μ observed by us, which may result from adsorbed OH. Although the O–H stretching vibration normally occurs at lower wavelengths (2.9μ), the OH bond is sufficiently weakened during adsorption so that the OH stretching band could be raised to 3.4μ (5).

The gaseous products were O_2, N_2, H_2O, and N_2O. We looked for H_2, but it was absent. The major gaseous products were O_2 and H_2O. Their relative amounts, compared with $[O_3]_0$, are listed in Table I for the unpacked reaction vessel. Their values do not vary with reactant pressures, and their average values and mean deviations are 1.05 $\pm$ 0.05 and 0.31 $\pm$ 0.03, respectively, for $[O_2]/[O_3]_0$ and $[H_2O]/[O_3]_0$. The minor products, N_2 and N_2O, deviate more. For $[NH_3]/[O_3]_0 < 100$, the relative values are reasonably constant and are 0.031 $\pm$ 0.008 and 0.032 $\pm$ 0.006, respectively, for $[N_2]/[O_3]_0$ and $[N_2O]/[O_3]_0$; for $[NH_3]/[O_3]_0 > 100$, both quantities are larger.

Table II.   The Reaction of O_3 with NH_3 in the Packed Cell

$[NH_3]/[O_3]_0$	$[O_3]_0$, Torr	$[NH_3]$, Torr	Temp, °C	k, min^{-1}
1.6	0.84	1.3	30	0.147
2.0	0.42	0.8	30.5	0.189
7.5	0.44	3.3	31	0.224
7.9	0.78	6.1	31	0.205
8.1	0.26	2.1	31	0.242
8.4	0.78	6.5	31	0.210
9.3	1.68	15.7	30	0.220
27.3	0.60	16.3	30.5	0.230
48	0.79	37.6	31	0.262
146	0.80	118	31	0.485[a]

[a] 3/2-order rate constant. Units are Torr$^{-1/2}$ min^{-1}.

The first-order rate coefficients, k, listed in Tables I and II increase with the ratio $[NH_3]/[O_3]_0$ as shown in Figure 4 but otherwise do not depend upon $[NH_3]$ or $[O_3]_0$. For any value of $[NH_3]/[O_3]_0$ the coefficients are larger in the packed than in the unpacked cell, showing heterogeneous effects. k can be fitted to the expression (see Figure 5).

$$k^{-2} = \alpha + \beta[O_3]_0/[NH_3] \qquad (c)$$

where α and β are, respectively, 22 and 144 sq min for the unpacked cell and 16 and 38 sq min for the packed cell.

The 3/2-order coefficient k' drops from 0.77 to 0.24 Torr$^{-1/2}$ min^{-1} as [NH$_3$] increases from 23.3 to 128 Torr. At higher NH$_3$ pressures, k' increases with [NH$_3$]. k' at the two highest NH$_3$ pressures may increase as a result of the sizeable effect of the aerosol attenuation at high [NH$_3$]/[O$_3$]$_0$ and as a result of diffusion effects. The aerosol effect was so important in the packed cell (because of the larger rate coefficients) that only one reliable value for k' could be obtained, and its value of 4.85 Torr$^{-1/2}$ min^{-1} was larger than in the unpacked cell for similar [NH$_3$]/[O$_3$]$_0$ ratios.

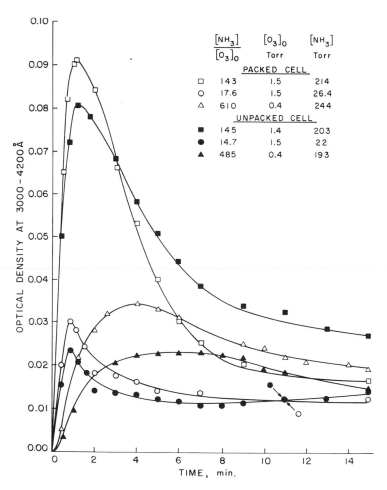

Figure 3.   Plot of the change in optical density at 3000–4200A (mainly 3660A) vs. reaction time for several runs in the unpacked and packed reaction cell. At these wavelengths the reactants do not give measurable absorption. The change in optical density is caused by aerosol formation.

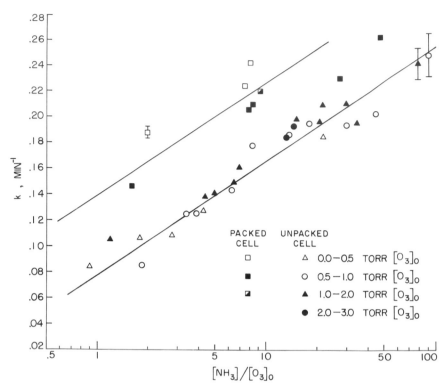

Figure 4. Semilog plots of the first-order rate constant, k, vs. $[NH_3]/[O_3]_0$ in the unpacked and packed reaction cell

Some runs were done with other gases added (*see* Table III); two runs were done with the major product, O_2, added to see if it influenced the reaction. The amount of O_2 was comparable with that produced in a run. It did not affect the order of the reaction, the value of the rate coefficient, k, or the relative amounts of the other gaseous products.

Two runs were done with H_2 added. One was with a small amount of H_2, which did not influence the reaction, to see if the H_2 would be recovered. It was completely recovered after the run, showing that our finding of no H_2 production was correct; any H_2 would have been detected if formed. The other run had 850μ of H_2 added. In this case some of the H_2 was consumed, and the production of O_2 was enhanced.

Several experiments were done in the unpacked and packed vessel in the presence of a large excess of He to test for inert gas effects. The $[NH_3]/[O_3]_0$ ratios varied from 1.5–23, and all the decay curves obeyed first-order kinetics. The relative gaseous product yields were essentially unaffected in the unpacked cell (products were not analyzed in the packed cell), except possibly $[O_2]/[O_3]_0$ dropped slightly. However the rate coefficients k were reduced as the He pressure was raised, but the effect was more pronounced in the unpacked cell.

The results with He are graphically shown in Figure 6 where $(k_0/k)^2$ is plotted *vs.* [He]. In this plot k is the first-order decay coefficient with He present, and k_0 is the first-order decay coefficient for He absent as determined from Figure 5 at the same value of $[NH_3]/[O_3]_0$. The plot is quite linear for the unpacked cell and reasonably linear for the packed cell. The values of k_0/k are functions of [He] and not functions of $[He]/[NH_3]$, an unexpected observation (*See* Discussion). The relationship is

$$(k_0/k)^2 = 1 + \gamma[He] \qquad (d)$$

where γ is 0.0140 and 0.0055 Torr^{-1}, respectively, for the unpacked and packed cell.

Discussion

The results give a complex rate law but a simple dependence for the forming of products. The major gaseous products, O_2 and H_2O, are formed in the same relative amounts in each experiment, and N_2 and N_2O are reasonably independent of conditions, at least for $[NH_3]/[O_3]_0$

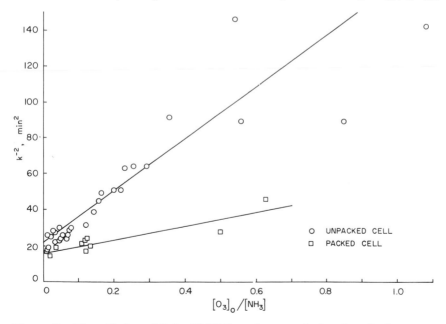

Figure 5.   Plot of k^{-2} *vs.* $[O_3]_0/[NH_3]$ *in the unpacked and packed reaction cell*

Table III.   The Reaction of O_3 with

[X], Torr	$[NH_3]/[O_3]_0$	$[O_3]_0$, Torr	$[NH_3]$, Torr	Temp, °C	k, min^{-1}
	X = O_2, Unpacked Cell				
1.30	13.6	0.93	12.8	29.4	0.152
2.10	20.1	1.62	32.6	29.0	0.180
	X = H_2, Unpacked Cell				
0.014 [a]	21.7	1.11	24.1	31.1	0.191
0.850 [b]	29	1.00	29	32.8	0.202
	X = He, Unpacked Cell				
66	16.7	1.48	24.9	29.2	0.151
161	22	0.71	16	28.4	0.105
380	~12	0.46	~5.5	27.9	0.064
445	23	0.72	16.9	26.5	0.063
546	15.6	1.52	23.8	28.2	0.066
785	15	0.81	12	29.0	0.048
1058	18	1.63	29.5	28.0	0.048
~1300	~20	0.41	~8	27.5	0.036
	X = He, Packed Cell				
180	7.9	0.43	3.4	—	0.136
195	8.8	0.81	7.1	—	0.135
225	9.6	1.60	15.4	—	0.140
815	~1.5	0.44	<0.8	—	0.074
824	8.8	0.74	6.5	—	0.091
844	8.0	0.44	3.5	—	0.090
855	22.2	0.46	10.2	—	0.110
880	9.5	1.64	15.5	—	0.099
900	23.0	1.64	37.7	—	0.110
925	7.6	1.71	12.9	—	0.094

[a] 100% of H_2 recovered at end of run.

< 100. The overall stoichiometry for the major products is consistent with one of the expressions of Strecker and Thienemann (3).

$$2NH_3 + 4O_3 \rightarrow 4O_2 + H_2O + NH_4NO_3$$

The O_2 produced was equal to, or slightly greater than, the O_3 consumed. This suggests that whenever an O_3 reacts, O_2 must be a product; this inference will greatly limit mechanistic possibilities. The above stoichiometric equation predicts $[H_2O]/[O_3]_0 = 0.25$, but we observed 0.31 ± 0.03. The additional H_2O is associated with the production of the minor products N_2 and N_2O.

The rate law is more complex. Two results give evidence for a free radical chain mechanism. The more compelling is the non-integral order of the reaction under some conditions—i.e., 3/2-order at $[NH_3]/$

NH₃ in the Presence of Added Gases

$[O_2]/[O_3]_0$	$[N_2]/[O_3]_0$	$[H_2O]/[O_3]_0$	$[N_2O]/[O_3]_0$
	$X = O_2$ Unpacked Cell		
—	0.028	—	—
—	0.040	0.31	0.026
	$X = H_2$, Unpacked Cell		
1.04	—	—	—
1.30	—	—	—
	$X = He$, Unpacked Cell		
0.94	0.028	—	—
0.85	0.030	0.38	0.038
1.00	0.043	0.35	0.039
0.87	0.042	0.30	0.034
0.85	0.030	0.34	0.020
0.96	0.024	0.35	0.030
1.08	0.044	0.33	0.019
1.07	0.065	—	0.035
Ave. = 0.95	0.038	0.34	0.031
±0.08	±0.010	±0.02	±0.007
	$X = He$, Packed Cell		
—	—	—	—
—	—	—	—
—	—	—	—
—	—	—	—
—	—	—	—
—	—	—	—
—	—	—	—
—	—	—	—
—	—	—	—
—	—	—	—

[b] 70% of H₂ recovered at end of run.

$[O_3] > 120$; the other evidence is the experiment with 850μ of H_2 added. Some of the H_2 was consumed, indicating an intermediate that readily attacks H_2 at room temperature; such an intermediate is the HO radical. Further evidence for HO is that $[O_2]/[O_3]_0$-is much greater in the presence of H_2. This occurs when HO attacks H_2 to produce H, which would then react with O_3 to produce excited HO which induces the chain decomposing of O_3 to produce O_2 (6).

Other results indicate the importance of wall reactions. First is that k depends on $[O_3]_0$. It is unusual for a rate coefficient to depend on the initial concentration of a reactant when that reactant is entirely consumed in the reaction. A similar result was found in the O_3–CS_2 reaction where wall reactions were also indicated (7). The importance of wall reactions was confirmed by the experiments in the packed reac-

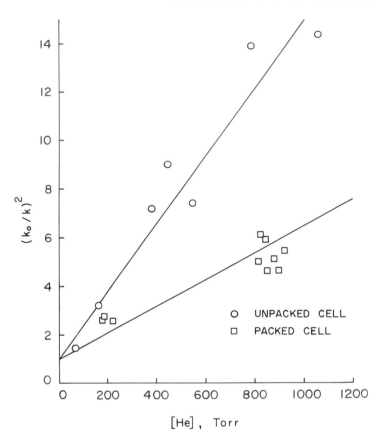

Figure 6. *Plot of the square of the ratio of the rate constant in the absence of He, k_0, to that in the presence of He, k, vs. the He pressure in the unpacked and packed reaction cell (the values for k_0 were taken from Figure 5 at the values of $[O_3]_0/[NH_3]$ corresponding to each k)*

tion vessel where the rate coefficients were larger for conditions comparable with those in the unpacked cell; under some conditions k was 60% larger.

Also supporting wall reactions is the experiments in excess He. He, an inert gas, reduces the rate coefficient which indicates that either the chain termination steps require a chaperone gas or that initiation is a diffusion-controlled wall reaction. Since the He effect is much more pronounced in the unpacked cell, the latter of the two possibilities is supported.

To formulate a mechanism, it is necessary to use a reaction scheme that,

    1. produces O_2 whenever O_3 reacts

2. gives a free radical chain mechanism with HO being the probable chain carrier

3. involves wall initiation

The initiating steps are probably those of the heterogeneous decay of O_3 which occurs even in the absence of NH_3. The predicted reaction steps are

$$2O_3 \quad \xrightarrow{\text{wall}} \quad 2O_2 + O_2{}^* \tag{1}$$

$$O_2{}^* + O_3 \rightarrow 2O_2 + O \tag{2}$$

$$O + O_3 \quad \rightarrow \quad 2O_2 \tag{3}$$

where $O_2{}^*$ is an excited O_2 molecule, possibly in a singlet state. Reaction 1 occurs on the wall and is highly exothermic. It is not unreasonable to expect that some of the O_2 produced is energized. The excited O_2 produced could easily have enough energy to dissociate another O_3 molecule. If singlet O_2 is produced, Reaction 2 is well established (8). The O atom would react with O_3 via Reaction 3 or in the presence of NH_3; it could also react via

$$O + NH_3 \rightarrow HO + NH_2 \tag{4}$$

When HO is produced, the chain is propagated by

$$HO + NH_3 \rightarrow H_2O + NH_2 \tag{5}$$

The NH_2 radical reacts readily with O_3, and since O_2 must be produced, the only exothermic reaction is

$$NH_2 + O_3 \rightarrow NH_2O + O_2 \tag{6}$$

The chain carrier, HO, is probably regenerated by the reaction

$$NH_2O + O_3 \rightarrow HO + O_2 + HNO \tag{7}$$

One complication is the reaction of NH_2 with O_2, a product of the reaction. Two possible reactions have been proposed for the NH_2–O_2 interaction (1)

$$NH_2 + O_2 \rightarrow H_2O + NO$$

or

$$NH_2 + O_2 \rightarrow HNO + HO$$

The former reaction is a chain terminating reaction and cannot be important in our system since the rate coefficient is unaffected even when 2.1 Torr of O_2 was added. The latter reaction regenerates the chain and produces HNO and thus is similar to Reaction 7; however, it consumes O_2. If it were as important as Reaction 6, $[O_2]/[O_3]_0$ should drop below unity. Either its rate constant is measurably smaller than that of Reaction 6, or the reaction proceeds through an intermediate complex which lives long enough to react with O_3, *viz*

$$NH_2 + O_2 \rightarrow NH_2O_2 \qquad (8)$$

$$NH_2O_2 + O_3 \rightarrow NH_2O + 2O_2 \qquad (9)$$

Reaction 8 followed by Reaction 9 is kinetically indistinguishable from Reaction 6.

Termination probably occurs from the interaction of two NH_2O radicals since NH_2 must be rapidly scavenged by either O_3 or O_2. The reaction is represented as

$$2NH_2O \rightarrow \text{Termination} \qquad (10)$$

The products of the reaction are unimportant and undetectable in our system because Reaction 10 occurs infrequently. An upper limit to Reaction 10's importance is made by considering the following information. The ratio k_3/k_4 is about 75, discussed below, but even at $[NH_3]/[O_3]_0$ $\sim$ 1, $[O_2]/[O_3]_0 \sim$ 1.0, which indicates that Reaction 4 leads ultimately to the bulk of O_3 decomposition. The chain lengths must be large—i.e., $\geqslant$ 500; also four O_3 molecules are consumed in each chain cycle. Therefore any product from Reaction 10 has a final concentration $< 5 \times 10^{-4}$ $[O_3]_0$, which would be undetectable. Nevertheless we can speculate about Reaction 10. For example, the isomeric HNOH form of NH_2O possibly is involved and reacts with itself to produce N_2 + $2H_2O$ via the intermediate

$$\begin{array}{ccc} H - O - N - H \\ \vdots \quad \vdots \quad \vdots \\ H - N - O - H \end{array}$$

The reaction might be concerted or could initially proceed by one four-center step in which one H_2O molecule was eliminated, followed by a second four-center step in which N_2 and the other H_2O molecule were produced.

The nitrogen-bearing products are produced from the ozonation of HNO. A possible sequence of steps is

$$HNO + O_3 \quad \rightarrow HNO_2 + O_2 \qquad (11)$$

$$HNO_2 + NH_3 \rightarrow NH_4NO_2 \qquad (12a)$$

$$\rightarrow N_2 + 2H_2O \qquad (12b)$$

$$NH_4NO_2 + O_3 \rightarrow NH_4NO_3 + O_2 \qquad (13a)$$

$$\rightarrow N_2O + 2H_2O + O_2 \qquad (13b)$$

where Reactions 12b and 13b are composite reactions. Alternately, HNO_2 could react with O_3

$$HNO_2 + O_3 \quad \rightarrow HNO_3 + O_2 \qquad (14)$$

$$HNO_3 + NH_3 \rightarrow NH_4NO_3 \qquad (15)$$

The latter sequence does not produce N_2 or N_2O. The competition between the two sequences may partially explain the increase in $[N_2]/[O_3]_0$ and $[N_2O]/[O_3]_0$ at large $[NH_3]/[O_3]_0$.

The reaction mechanism outlined above predicts that $[O_2]/[O_3]_0$ should drop from 1.5 in the absence of NH_3 to 1.0 for a long chain reaction in the presence of NH_3, which agrees well with the observations. The predicted value for $[H_2O]/[O_3]_0$ is $0.25 + (2.25[N_2] + 2[N_2O])/[O_3]_0$. The former quantity is 0.31 ± 0.03 in the absence of He and 0.34 ± 0.02 in the presence of He; whereas the latter quantity is 0.38 ± 0.03 in the absence of He and 0.40 ± 0.03 in the presence of He. In both cases, the $[H_2O]$ is low, probably because of losses by adsorption to the NH_4NO_3 solid or the walls of the vacuum line. However, the discrepancy is not outside the experimental uncertainty.

With the steady state hypothesis for the reactive intermediates, Reactions 1–15 lead to the rate law

$$\frac{-d[O_3]}{dt} = 4R_i + \frac{k_4[NH_3]R_i}{k_3[O_3] + k_4[NH_3]}$$

$$+ 4k_7[O_3]\left[\frac{k_4[NH_3]R_i/k_{10}}{k_3[O_3] + k_4[NH_3]}\right]^{1/2} \qquad (e)$$

where R_i is the rate of the initiation step, Reaction 1. In the above derivation Reaction 12b has been neglected. Including it complicates the expression and only reduces $-d[O_3]/dt$ by about 1%. If the chain length is long, which is so since $[O_2]/[O_3]_0 \sim 1.0$, then Equation e simplifies to

$$-d[O_3]/dt \simeq 4k_7[O_3]\left[\frac{k_4[NH_3]R_i/k_{10}}{k_3[O_3] + k_4[NH_3]}\right]^{1/2} \qquad (f)$$

The expression for R_i is not obvious since the initiating reaction is a diffusion-inhibited wall reaction. Empirically, the form needed for R_i to fit the experimental observations for k is

$$R_i = k_1[O_3]/([O_3]_0 + \alpha[NH_3]/\beta)(1 + \gamma[He]) \qquad (g)$$

with $\beta = k_3k_{10}/16k_1k_4k_7^2$. In the absence of He and at low values of $[NH_3]/[O_3]_0$, Equation g reduces to

$$R_i \simeq k_1[O_3]/[O_3]_0 \qquad (h)$$

This relationship suggests that the wall rate for some reason depends inversely on $[O_3]_0$. The rate constant k_1 in the packed cell, relative to that in the unpacked cell, is given by the inverse ratio of the corresponding slopes in Figure 5. This value is 3.8 which compares with the increase in surface to volume ratio of a factor of 3.4 when the cell is packed. As the $[NH_3]/[O_3]_0$ ratio is increased, the relative importance of the interaction of NH_3 with the surface increases, and the initiating rate law shifts to

$$R_i \simeq k_1[O_3]\beta/\alpha[NH_3] \qquad \text{(i)}$$

The ratio α/β is the ratio of intercept to slope in Figure 5 for the packed and unpacked cell. Since this ratio is larger in the packed cell, the influence of increasing $[NH_3]/[O_3]_0$ is felt sooner in the packed cell.

With He added the other term in Equation g reflects diffusional inhibition. This effect is more pronounced in the unpacked cell as would be expected. However, the form of the term suggests that the diffusional effect is for NH_3, rather than O_3, diffusing to the wall. Otherwise the term should be $(1 + \gamma[He] + \gamma'[NH_3])$.

Regardless of the complexities of the wall initiation step, which are not well understood, Equation g can be substituted into Equation f to give a generalized rate law for long chains

$$\frac{-d[O_3]}{dt} \simeq \qquad \text{(j)}$$

$$4k_7[O_3]\left[\frac{k_1k_4[NH_3][O_3]/k_{10}}{(k_3[O_3] + k_4[NH_3])([O_3]_0 + \alpha[NH_3]/\beta)(1 + \gamma[He])}\right]^{1/2}$$

The first prediction from Equation j is that the O_3 decay is first-order for $k_3[O_3] > k_4[NH_3]$ and 3/2-order for $k_3[O_3] < k_4[NH_3]$. We have observed both rate laws and found that the cross-over occurs for $[NH_3]/[O_3] \sim 100$; consequently, k_3/k_4 should be ~ 100. The rate constants k_3 and k_4 have been measured. We believe the best value (9) for k_3 to be $5.2 \times 10^6 M^{-1}$ sec^{-1} at 30°C. The two most reliable results for k_4 are given by Albers et al. (10) and Kurylo et al. (11). Both of their Arrhenius expressions give the same value of $7.0 \times 10^4 M^{-1}$ sec^{-1} for k_4 at 30°C. The ratio of these reported rate constants is 75, which supports our conclusion.

It is a straightforward procedure to show that when $k_3[O_3] >> k_4[NH_3]$ Equation j reduces to a first-order rate law in $[O_3]$ with a rate constant which conforms to Equation c in the absence of He and Equation d in the presence of He. We forced the form of R_i for this to be true; however for $k_4[NH_3] >> k_3[O_3]$ and He absent, Equation j reduces to

$$-d[O_3]/dt \simeq 4k_7(k_1\beta/\alpha k_{10}[NH_3])^{1/2}[O_3]^{3/2} \qquad \text{(k)}$$

This rate law is 3/2-order in $[O_3]$ as experimentally found and predicts that the 3/2-order rate constants, k', should vary as $[NH_3]^{-1/2}$.

Figure 7 is a plot of $(k')^{-2}$ vs. $[NH_3]$ in the unpacked cell. The two values at the two highest NH_3 pressures were omitted since they do not fit on the plot. The remaining data are scattered but they fit a straight line which passes through the origin. Some of the scatter can be attributed to variation in the temperature among the runs. The slope of the line is 0.13 sq min. The ratio of this value to the intercept of Figure 5 (22 sq min) gives $k_4/k_3 = 0.006$, which is about a factor of two lower than the ratio of 0.013 expected from literature values (*See* previous discussion).

The reason for the discrepancy, as well as the failure of the two points at highest NH_3 pressure to fit the plot, is uncertain, but the complications at the high NH_3 pressures are severe:

1. The aerosol interferes with experimentally determining k' and probably accounts for the scatter.

2. The aerosol particles in the gas phase can become centers for the heterogeneous reaction.

3. NH_3 reacts readily with the surface at high NH_3 pressures and thus continuously changes its nature throughout the run.

4. The diffusion effects become pronounced so that concentrations at the surfaces are different from those in the gas. This gradient in concentration not only affects the reaction rate but also affects the measured O_3 concentration.

Therefore the rate law is more complex than indicated by Equation k.

Another complication is introduced at high pressures by the reaction

$$O + O_2 + M \rightarrow O_3 + M$$

which has a rate constant (*12*) of $2 \times 10^8 M^{-2}$ sec^{-1}. This reaction should

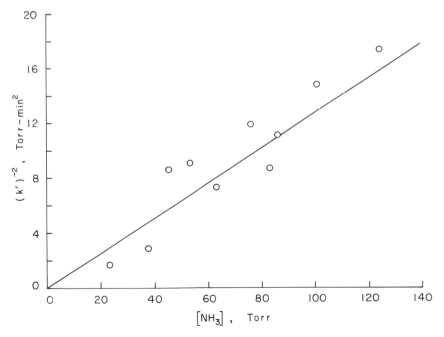

Figure 7.   Plot of $(k')^{-2}$ vs. $[NH_3]$ for runs with $[NH_3]/[O_3]_0$ between 44–130 for the unpacked reaction cell

inhibit the rate as the reaction proceeds, the O_3 becomes consumed, and O_2 accumulates. The inhibition should have been apparent in the presence of excess He, but it was not.

To prevent this difficulty, the reaction mechanism can be slightly altered if the O_2^* produced in Reaction 1 is assumed to be vibrationally excited O_2. Then Reactions 2, 3, and 4 are replaced by

$$O_2^* + O_3 \rightarrow O_2 + O_3 \tag{3'}$$

$$O_2^* + NH_3 \rightarrow HO + NH_2O \tag{4'}$$

The rate expression, Equation e, would become

$$\frac{-d[O_3]}{dt} = 3R_i + 4k_7[O_3] \left[\frac{k_4'[NH_3]R_i/k_{10}}{k_3'[O_3] + k_4'[NH_3]} \right]^{1/2} \tag{e'}$$

and the rest of the analysis would be unchanged. However, since the reaction order with respect to $[O_3]$ is not changed by adding He, this mechanism would require that He be particularly inefficient in quenching O_2^*. This alternative mechanism also has shortcomings because it is not clear why He should be so inefficient. It is clear that initiation steps in the O_3–NH_3 reaction are not well understood.

Acknowledgment

We wish to thank E. Lissi and W. B. DeMore for useful suggestions. This work was supported by the Environmental Protection Agency through the Air Pollution Control Organization under Grant No. AP 00022, for which we are grateful.

Literature Cited

1. Cohen, N., Heicklen, J., "Comprehensive Chemical Kinetics," Vol. 6, Elsevier, in press.
2. Bacon, H. E., Duncan, A. B. F., *J. Amer. Chem. Soc.* (1934) **56,** 336.
3. Strecker, W., Thienemann, H., *Ber. Chem. Gesell* (1920) **53,** 2098.
4. Silverstein, R. M., Bassler, G. C., "Spectrometric Identification of Organic Compounds," p. 99, Wiley, New York, 1967.
5. Cusumano, J. A., Low, M. J. D., *J. Phys. Chem.* (1970) **74,** 1950.
6. McGrath, W. D., Norrish, R. G. W., *Proc. Roy. Soc. (London)* (1960) **A254,** 317.
7. Olszyna, K. J., Heicklen, J., *J. Phys. Chem.* (1970) **74,** 4188.
8. Zipf, E. C., *Can. J. Chem.* (1969) **47,** 1863.
9. Krezenski, D. C., Simonaitis, R., Heicklen, J., *Int. J. Chem. Kinetics* (1971) **3,** 467.
10. Albers, E. A., Hoyermann, K., Wagner, H. Gg., Wolfrum, J., *Symp. Intern. Combust., 12th* (1969) 313.
11. Kurylo, M. J., Hollinden, G. A., LeFevre, H. F., Timmons, R. B., *J. Chem. Phys.* (1969) **51,** 4497.
12. Benson, S. W., DeMore, W. B., *Ann. Rev. Phys. Chem.* (1965) **16,** 397.

RECEIVED May 10, 1971.

Formation and Destruction of Ozone in a Simulated Natural System (Nitrogen Dioxide + α-Pinene + hν)

DANIEL LILLIAN

Department of Environmental Sciences, Rutgers, The State University, New Brunswick, N. J.

The hypothesis that the naturally occurring system NO₂ + α-pinene + hν (3000–4000 A), would show analogous behavior to simplified systems of photochemical smog was tested statistically and verified. During the reaction ozone and organic oxidants were formed and consumed; aerosol (condensation nuclei) was formed. These data indicate that the naturally occurring photo-oxidation of α-pinene may serve as a sink for the significant quantities of terpenes that are emitted globally, a source and sink for ozone, and a source of the naturally occurring light scattering aerosol (blue haze).

Trying to account for the fate of an estimated 10^8 tons of terpenes emitted annually from plants, Went (1) hypothesized that terpenes underwent reactions similar to those of olefins in photochemical smog. He suggested that the photo-oxidation of the terpenes was responsible for forming the blue haze observed over densely vegetated areas. Observing a Tyndall beam when ozone was allowed to react with a terpene (1) and detecting condensation nuclei when a blend of NO_2 and α-pinene was irradiated with sunlight (2) supported this hypothesis.

Here, Went's hypothesis, that the naturally occuring system NO_2 + α-pinene + hν behaves analogously to simplified models of photochemical smog is tested. Besides suggesting the mode of natural aerosol formation, the proof of this hypothesis has important implications in the atmospheric chemistry of other nonurban trace constituents, particularly ozone.

Experimental

Methods. Blends of 10 pphm nitrogen dioxide and 50 pphm α-pinene in dry air (absolute humidity 0.0005 gram H_2O/gram dry air) were prepared in 150-liter transparent bags and irradiated at 25°C for 120 minutes. All experiments were performed in duplicate and blanks were run according to a complete factorial design. Statistical comparisons were made of the mean value of a given variable with the mean value of the appropriate blank using Tukey's (*3*) method of multiple comparisons. The results are reported at a .05 confidence level.

During the course of the irradiations, the concentrations of the following variables were monitored: oxidants (Mast coulometric ozone meter) (*4*), condensation nuclei with radii greater than 10^{-7} cm (G.E. Type CN small particle detector) (*5*), ozone (Regener chemiluminescent ozone meter (*6*), nitrogen dioxide and nitric oxide (Saltzman method) (*7*), and α-pinene (Perkin Elmer model 800 gas chromatograph).

The gas chromatograph was equipped with a flame ionization detector. A 50-foot length of 0.020 inch i.d. stainless steel open tubular capillary column coated with Carbowax 1540 served as the main column. A freeze out trapping technique was used to concentrate the α-pinene before entering the main column. The pre-column trap consisted of an in-line capillary column, identical to the main one, inserted between the injector and inlet of the main column. The trap was located outside the oven and cooled with a dry ice–ethanol bath before injection of a 5 cc sample. A 80°C hot water bath was used to release the α-pinene. The operating conditions of the gas chromatograph were as follows:

Helium (carrier) flow rate 3.1 cc/minute

Hydrogen flow rate 50 cc/minute

Air flow rate 850 cc/minute

Helium (make up) flow rate 40 cc/minute

Oven temperature (isothermal) 100°C

Reaction Bags and Irradiation Chamber. Teflon (50-mil FEP Type C) bags of 32 inches × 48 inches were fabricated by impulse heat sealing. The bags were fitted with glass ball joints to connect them to the TFE Teflon sampling lines.

A Hotpack controlled environmental room was used as an irradiation and constant temperature chamber. By maintaining the chamber temperature at 21 ± 1°C, a temperature of 25 ± 2°C was achieved in the Teflon reaction bag.

Radiation simulating solar radiation was given by a bank of four G.E. 40-watt cool white fluorescent lamps and two Westinghouse 400-watt EH1 mercury vapor lamps mounted on one wall of the chamber. The walls of the chamber were covered with aluminized Mylar which provided a reflecting surface. ϕk_a for NO_2 was 2.8 hr^{-1}.

Chemicals Used. Listed below are the specifications of the chemicals used in this study. Common laboratory reagents used for the various standard analyses met with the specifications prescribed in the cited methods and are not listed. All gases were supplied by the Matheson Company, East Rutherford, N. J.

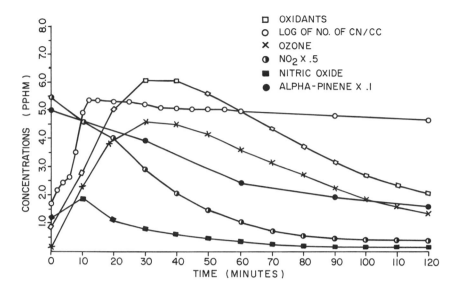

Figure 1.   Concentration–time profiles of the indicated variables for the system
$NO_2 + \alpha\text{-pinene} + h\nu$

Air, zero gas: maximum dew point −78°F, less than 0.5 ppm hydro-carbon equivalent to methane

Hydrogen: pre-purified grade, 99.95% minimum purity

Nitrogen dioxide: 99.5% minimum purity

Helium: ultra-high purity grade, minimum purity 99.999%

Oxygen: extra-dry grade, maximum dew point −79°F, minimum purity 99.6%

α-Pinene: minimum purity 99.9%, supplied by Glidden Chemical Company, Jacksonville, Fla.

Results

The concentration–time profiles obtained upon irradiating synthetic blends of nitrogen dioxide and α-pinene in air, absolute humidity <0.0005 gram H_2O/gram dry air (the effect of water vapor on the system $NO_2 +$ α-pinene + hν has already been reported (8)), are given for total oxidants, condensation nuclei, ozone, nitrogen dioxide, nitric oxide, and α-pinene in Figure 1. The main effects of irradiation are consumption of NO_2 and α-pinene and production of ozone, organic oxidants (the differ-ence between total oxidants and ozone), and condensation nucli. By analogy to simplified chemical models of photochemical smog (9, 10, 11, 12, 13), these effects and the associated profiles result from free radical reactions initiated by electrophillic attacks of atomic oxygen and/or ozone

—depending on the respective reaction rate constants and concentrations of the two (14)—at the α-pinene double bond.

By analogy to Cvetanovic's proposed mechanism for the reaction of atomic oxygen with simple olefins (9), one would expect a diradical to form from the reaction of atomic oxygen with α-pinene. However the cyclic structure of the α-pinene and the strained four-membered ring preclude assigning a structure to this intermediate pending experimental evidence. Similarly the intermediate formed by the ozone–α-pinene reaction is not amenable to a rigorous comparison with the product of the π-complex proposed in Criegee's zwitterion mechanism for simple olefin–ozone reactions (9).

Ozone and Organic Oxidants. When a blend of 10 pphm NO_2 in air was irradiated under the same experimental conditions used to obtain the data of Figure 1, O_3 and NO increased to about 1.2 pphm within the first five minutes of irradiation. Their concentrations remained at that value during the 120-minute experiment. This buildup and attainment of steady state is attributed to the rapid synthesis of NO and ozone, according to Reactions 1 and 2, and at equilibrium to the equally rapid removal of NO and O_3 by Equation 3:

$$NO_2 + h\nu \rightarrow NO + O \tag{1}$$

$$O + O_2 + M \rightarrow O_3 + M \tag{2}$$

$$O_3 + NO \rightarrow NO_2 + O_2 \tag{3}$$

$$\frac{\phi k_a}{k_3} = \frac{[NO][O_3]}{[NO_2]} \tag{4}$$

M represents a third body—$e.g.$, nitrogen, ϕ is the quantum yield for NO_2 photolysis, k_a is its specific absorption rate, and k_3 is the bimolecular reaction rate constant for Reaction 3.

Since Reactions 1, 2, and 3 are much faster than competing reactions involving olefins, the equilibrium relationship 4 must hold even in the presence of α-pinene (15). The buildup of ozone in Figure 1 above the steady state concentration it shows when no α-pinene is present is therefore accompanied by an increase in the NO_2:NO ratio. This increase is effected by reactions which convert NO to NO_2 as shown by a few reactions from Wayne's (11) mechanism for the NO_2-initiated photo-oxidation of an olefin:

$$\alpha p + O \rightarrow \alpha pO^* \tag{5}$$

$$\alpha pO^* + O_2 \rightarrow \alpha pO_3^* \tag{6}$$

$$\alpha p + O_3 \rightarrow \alpha pO_3^* \tag{7}$$

$$\alpha pO_3{}^* \to \text{Aldehydes}, + \text{RO} \cdot \; \dot{+} \; R\dot{C}O \tag{8}$$

$$\text{RO} \cdot \; + \text{NO} + O_2 \to RO_2 + NO_2 \tag{9}$$

$$RO_2 + \text{NO} \to \text{RO} + NO_2 \tag{10}$$

αp refers to α-pinene and the asterisk designates an unstable intermediate. As pointed out by Leighton (9), synthesis of ozone by free radical reactions with molecular oxygen may similarly lead to a buildup of ozone above steady state.

$$RO_2 + O_2 \to \text{RO} \cdot \; + O_3 \tag{11}$$

The difference between simultaneous oxidant and ozone readings (Figure 1) are attributed to organic oxidants formed from reactions of free radicals, the oxides of nitrogen, and the allotropes of oxygen. The positive 10% response of the Mast instrument to NO_2 accounts for only a small part of this difference, particularly in the latter stages of irradiation when the NO_2 concentration is low. The nature of the organic oxidants is speculative. However Stephens observed that the system NO_2 + α-pinene + hν formed PAN (16), indicating that part of the organic oxidant is attributable to this well known lachrymator and phytotoxicant.

Condensation Nuclei. Many mechanisms have been proposed (9) involving free radical polymerizations of various radicals which could lead to formation of condensation nuclei. It seems that if condensation nuclei are formed by such reactions, the myriad different radicals in a given system would lead to formation of a highly mixed polymer. Noting this, an oversimplified mechanism by which the system NO_2 + α-pinene + hν may form condensation nuclei (Figure 1) is for example, reactions of the alkyl peroxy radical formed in Reaction 9 with α-pinene and molecular oxygen:

$$\text{ROO} \cdot \; + \alpha p \qquad \qquad \text{ROO} \; \alpha p \cdot \tag{12}$$

$$\text{ROO} \; \alpha p \cdot \; + O_2 \qquad \qquad \text{ROO} \; \alpha pOO \cdot \tag{13}$$

$$\text{ROO} \; \alpha pOO \cdot \; + \alpha p \qquad \qquad \text{ROO} \; \alpha pOO \; \alpha p \cdot \tag{14}$$

The leveling off of the condensation nuclei concentration after the first few minutes of the irradiation indicates that the size distribution is shifting to larger particles as oxygenated olefin is incorporated into the aerocolloidal mass.

Ripperton *et al.* (17) and Groblicki and Nebel (18) have shown that the dark-phase reaction of ozone and α-pinene leads to rapid formation of condensation nuclei. Since relatively high concentrations of ozone are produced by the photochemical system NO_2 + α-pinene + hν, the ozone–

α-pinene reaction probably is responsible for a significant fraction of the condensation nuclei observed here. The relative importance of this mechanism of condensation nuclei production to one involving a peroxy free radical polymerization initiated by an atomic-oxygen–α-pinene reaction, however, cannot be assessed from the data available.

Using experimental techniques and conditions identical to those used for the controlled irradiations of gaseous mixtures in Teflon bags, a fifteen minute irradiation of a blend of 1 ppm NO_2 and 1 ppm α-pinene yielded a barely perceptible bluish haze in a Tyndall beam. Adding more ozone with a six-inch ultraviolet (uv) Penray lamp fitted into the top ball-joint of the 50-liter flask intensified this haze within seconds. Figure 2 is a picture of the Tyndall beam taken two minutes after the uv lamp had been activated for ten seconds. Over a substantially longer path length, the system $NO_2 + \alpha$-pinene $+ h\nu$ at concentrations near natural concentrations should be capable of forming the haze observed over densely vegetated areas (blue haze).

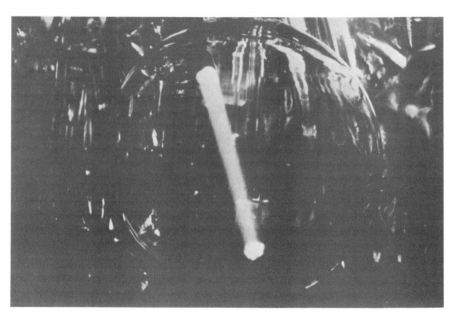

Figure 2.   Light scattering aerosol as viewed in a Tyndall beam

Nitric Oxide. Since photolysis of NO_2 did not proceed measurably before irradiation, as indicated by zero ozone readings for systems containing NO_2 in zero air, the NO readings obtained before irradiation for these systems and for the system of Figure 1 are artifacts of the analytical method. The positive error is probably attributable to the scrubbing column used to remove NO_2 before the NO oxidation step. Either less than 100% absorption efficiency was obtained, or nitric oxide was synthesized in the Saltzman scrubbing reagent (*19*):

$$3NO_2 + H_2O \rightarrow 2HNO_3 + NO$$

Both artifacts may have been concurrently operative.

Conclusion

The data presented have important implications in the behavior of tropospheric nonanthropogenic ozone, aerosol, and other trace constituents. Observational and experimental data have been reported by Ripperton *et al.* (*20*) indicating the natural synthesis of ozone in the troposphere. Considering this study, the ubiquitous presence of various terpenes (*21*), isoprene (*22*), and oxides of nitrogen (*20*) suggest that some ozone is synthesized in the lower troposphere by the reaction NO_2 + α-pinene + hν. Conversely, the destruction of ozone in the troposphere is partially ascribed to reactions with the terpenes and intermediates of the photochemical mixture.

During the photooxidative reactions, aerosols are formed which are undoubtedly similar to the aerosols forming the blue haze over densely vegetated areas. This aerosol may also account for a significant quantity of the natural organic continental aerosol.

There is basically little difference between the mechanism of photochemical smog formation and the naturally occurring photo-oxidation of terpenes. The former, associated with a greater emission intensity of ozone precursors, however, leads to higher concentrations of those intermediates responsible for the undesirable effects of photochemical air pollution.

Literature Cited

1. Went, F. W., *Proc. Nat. Acad. Sci.* (1960) **46**, 212.
2. Went, F. W., *Tellus* (1966) **28**, 549.
3. Scheffe, H., "The Analysis of Variance," Wiley, New York, 1963.
4. Mast, G. H., Sanders, H. E., *I.S.A. Trans.* (1962) **1**, 325.
5. Rich, T. A., *Goefis. Pura, Appl.* (1955) **31**, 60.
6. Regener, V. H., *J. Geophys. Res.* (1964) **69**, 3795.
7. U. S. Public Health Service, "Selected Methods for the Measurement of Air Pollutants," U. S. Government Printing Office, Washington, D. C., 1965.
8. Ripperton, L. A., Lillian, D., 63rd Meeting, Air Pollution Control Association, Saint Louis (June, 1970).
9. Leighton, P. A., "Photochemistry of Air Pollution," Academic Press, New York, 1961.
10. Altshuller, A. P., Bufalini, J. J., *Photochem. Photobiol.* (1965) **4**, 97.

11. Haagen-Smith, A. J., Wayne, G., "Air Pollution," A. C. Stern, Ed., Vol. 1, Academic, New York, 1968.
12. Pitts, J. N., Jr., "Photochemical Air Pollution: Singlet Molecular Oxygen as an Environmental Oxidant," *Advan. Environ. Sci.* (1969).
13. Altshuller, A. P., Bufalini, J. J., *Environ. Sci. Technol.* (1971) 1, 39.
14. Jaffee, S., Loudon, R., *Advan. Chem. Ser.* (1972) 113, 264.
15. Shuck, E. A., Stephens, E. R., "Oxides of Nitrogen," J. N. Pitts and R. L. Metcalf, Eds., *Advan. Environ. Sci.*, Wiley, New York, 1969.
16. Stephens, E. R., *Proc. Amer. Petrol. Inst.* (1962) 42, 665.
17. Ripperton, L. A., Jeffries, H. E., White, O., *Advan. Chem. Ser.* (1972) 113, 219.
18. Groblicki, P. J., Nebel, G. J., "The Photochemical Formation of Aerosols in Urban Atmospheres," General Motors Research Symposium on Chemical Reactions in Urban Atmospheres, Warren, Michigan (October, 1969).
19. Mueller, P. K., Transah, N. O., Tokiwa, Y., Kothny, E. L., "Series *vs.* Parallel Continuous Analysis for NO, NO_2, and NO_x. II. Laboratory Data," 9th Conference, Methods in Air Pollution and Industrial Hygiene Studies, Pasadena (February, 1968).
20. Ripperton, L. A., Jeffries, H., Worth, J. J. B., *Environ. Sci. Technol.* (1971) 5, 246.
21. Rasmussen, R. A., Went, F. W., *Proc. Nat. Acad. Sci.* (1965) 53, 215.
22. Rasmussen, R. A., *Environ. Sci. Technol.* (1970) 4, 667.

RECEIVED May 10, 1971.

8

Formation of Aerosols by Reaction of Ozone with Selected Hydrocarbons

L. A. RIPPERTON and H. E. JEFFRIES

Department of Environmental Sciences and Engineering, University of North Carolina, Chapel Hill, N. C. 27514

O. WHITE

Atomic Energy Commission, New York, N. Y. 10014

Dark phase reaction of ozone (O_3) with open chain mono-olefins in air produced no light-scattering aerosol detectable with the Goetz Moving Slide Impactor. Reactions of O_3 with the cyclic olefins, cyclohexene and α-pinene, and the diolefin, 1,5-hexadiene (reactants in the low or fractional parts per million concentration) produced detectable quantities of light-scattering aerosols. Aerosol generation was enhanced by increasing concentrations of either reactant. Reducing oxygen from 20 to 2% reduced aerosol generation; increasing water vapor concentration (~0.0, ~45, ~100% relative humidity, 21°C) enhanced aerosol generation. Kinetic data from the α-pinene–O_3 system suggest that α-pinene reacted with a product of the original reaction. This co-polymerization is proposed as an important step in forming organic particulate matter.

The role of olefin–ozone (O_3) reactions in generating aerosols from organic vapors and the influence of oxygen (O_2) and water vapor on aerosol production in these reactions have been studied here. Kinetic data were examined to try to derive a mechanism for forming aerosols from organic vapors.

Past literature shows that organic compounds are important in the formation and composition of aerosols (1, 2, 3), but this information has not been widely accepted or used until recently. In 1963 Junge (4) included only a few sentences on the organic content of aerosols; however,

Goetz (5), writing about maritime haze particles, said that the degrading of particles by ultraviolet radiation "appears to be definite proof for the significant, if not occasional dominant, presence of various hydrocarbons in the aerocolloidal matter."

How organic vapors are incorporated into the aerocolloidal mass is important in urban and in non-urban air. Polymeric material is generated when O_3 reacts with liquid olefins, and aerosols also result from the reaction of O_3 with gas phase olefins at pressures in the millimeter range. This led to the belief that photochemical haze actually resulted from O_3 reacting with olefins. Leighton (6) stated that early workers in air pollution chemistry found that olefins reacting with O_3 in the low parts per million (ppm) concentrations did not produce aerosols. In contrast, Prager et al. (7) found that cyclopentene, cyclohexene, or 1,5-hexadiene in the ppm range reacting with O_3 produce aerosols (hydrocarbon 10 ppm, O_3 5 ppm). Photochemical systems containing NO_2 and cyclic olefins or diolefins also produced aerosols.

Studying O_3 behavior in non-urban air, we followed up work of Went and Rasmussen (8) by considering the terpenes as possible gas phase destructive agents for atomspheric O_3—i.e., sinks for natural O_3. α-Pinene is the most abundant terpene found in the North Carolina pines (9, 10) and was the first terpene tested. No rate constant for the O_3–α-pinene reaction was found in the literature; therefore, a study was made to determine it. The results of determining the utilization rate of the reactants suggested strongly that aerocolloidal material was produced. Went (11) reported that the reaction yielded an aerosol which was formed rapidly. With reactants in the tens of parts per hundred million (pphm) range, we generated enough aerosol to produce a Tyndall beam in dark phase O_3–α-pinene reactions and in photochemical secondary

Table I. Aerosol Production from Selected Hydrocarbon–Ozone Systems[a]

Hydrocarbon Compound	Hydrocarbon Concentration ppm	Ozone ppm	Relative Light Scatter	
			In Air	In Nitrogen (2% Oxygen)
α-Pinene	1	0.60	2650	—
α-Pinene	1.5	0.14	700	90
1,5-Hexadiene	2.0	0.14	20	—
1,5-Hexadiene	8.0	3.00	5000	—
1,5-Hexadiene	2.0	0.40	—	70
Cyclohexane	8.0	0.58	0	—
2-Hexene	2	0.35	0	—

[a] Relative humidity = 0.01% at 21C°.

reactions of the system NO_2 + α-pinene + $h\nu$ (*12, 13, 14*). Placing freshly crushed pine needles in the presence of O_3 also generated aerosol.

In 1969 Groblicki (*15*) reported that aerosol formed in numerous O_3–olefin (dark phase) and NO_2–olefin (photochemical) systems in the absence of SO_2. He found little aerosol formed by reacting O_3 with open-chained olefins such as 1-heptene but found much aerosol formed when O_3 reacted with α-pinene and with cyclohexene in low ppm of reactant concentration (4 ppm hydrocarbon). He found that cyclopentene formed moderate amounts of aerosol and that O_3 concentration seemed to be a limiting factor in forming aerosol. From Groblicki's work and our study of the O_3–α-pinene reaction, we believed that we had insight as to a mechanism of organic aerosol formation. A key step in the process seems to be the forming of an intermediate diactive species.

Recent publications by Altshuller (*16*), Mueller *et al.* (*17*), and Robinson and Robbins (*18*) study the more general aspects of aerosols in the ambient air.

Table II.   Effect of Cyclohexene Concentration on Aerosol Production

Reactants		
Cyclohexene, ppm	Ozone, ppm	Relative Light Scatter[a]
0	0.60	0
0.5	0.60	$132 \begin{cases} 119 \\ 146 \end{cases}$
1.0	0.60	$521 \begin{cases} 500 \\ 543 \end{cases}$
1.6	0.60	1025
2.4	0.60	1523
3.2	0.60	1453

[a] Air, RH = 0.1%, 21°C.

Experimental

Generating Aerosols by Dark Phase Ozone Reaction.   Reactants, various hydrocarbons and O_3, were brought together in low ppm or tenths of a ppm concentration in 150-liter Teflon bags. The organic reactants used were α-pinene, cyclohexene, 1,5-hexadiene, cyclohexane, and 2-hexene. In all but one series of experiments, relative humidity was virtually zero (dew point < −48°C). In experiments using cyclohexene and O_3, the water vapor pressure was varied; relative humidities of ∼0.0%, ∼50%, and 95–100% were used at 21°C. α-Pinene and other hydrocarbons alone in clean air in Teflon bags do not deteriorate detectably in 24 hours; ozone deteriorates less than 5% in 18 hours.

Table III.   Aerosol Production: Cyclohexene and Ozone in

Reactants

Cyclohexene, ppm	Ozone, ppm
1	0
1	0.14
1	0.33
1	0.58
1	0.90

[a] Average values.
[b] Temperature: 21°C.

Generating Aerosols with Photochemical Systems. Systems containing cyclohexanone alone, cyclohexanone and 1-hexene, toluene and NO_2, 1,5-hexadiene alone, and 1,5-hexadiene and NO_2 (concentrations given in Table IV) were exposed to midday February sunlight for 1 hour in 150-liter Teflon bags and tested for aerosol generation with the Goetz Moving Slide Impactor.

α-Pinene and NO_2 (50 and 10 pphm, respectively) in 150-liter Teflon bags and 1-hexene and NO_2 (4 ppm each) in 200-liter Mylar bags were exposed to artificial light at about one-tenth noonday intensity and tested for the presence of condensation nuclei with the Gardner condensation nuclei counter.

Kinetic Studies. α-Pinene reacted with O_3 in Teflon or Mylar bags, and both reactants were analyzed. Data are presented in Figures 1, 2, and 3; kinetic information is included in Table V.

Analytical Techniques. Ozone was measured with a chemiluminescent O_3 meter of the type developed by Regener (19). α-Pinene was followed with a Perkin-Elmer Model 80 gas chromatograph equipped with a flame ionization detector. The columns were borosilicate glass, 12 feet long with a 3 mm inside diameter, produced by the authors with 80/90 Anakrom SD with a 4% liquid loading of Carbowax 20 M.

The aerosols were deposited, after reacting 1 hour, on a metalized slide with the Moving Slide Impactor (MSI) developed by Goetz (5). According to Goetz, the device is about 50% efficient for collecting particles of 0.1μ diameter, 80% for 0.2μ diameter, and essentially 100% for 0.3μ diameter and over. The flow rate through the instrument was 4 liters per minute, and the aerosol was deposited in 48 seconds over an area of 2×4 mm. The deposited material was analyzed by low angle (20°) dark field vertical illumination of the microscope slide where the material was deposited. Light-scattering units are arbitrary but represent nanoamperes registered on a light meter attached to the microscope.

Air and Nitrogen Atmospheres

Relative Light Scatter[b]

Air,[a] RH = 0.01%		Air, RH = 50%	Air, RH = 95-100%	Nitrogen, (2% Oxygen) RH = 0.01%
0		0	0	0
55	{39, 45, 81}	—	—	—
246	{195, 242, 300}	1150	2300	40
521	{500, 543}	1400	6400	156
706		—	—	421

To obtain infrared spectra of the depositable fraction of the particles, the particles were deposited on the crystal of an internal reflectance attachment to an infrared spectrophotometer using MSI. The IR spectra were then taken with a Perkin-Elmer model 337 grating infrared spectrophotometer.

Discussion of Results

The quantitative data on the production of light-scattering aerosols are given in Tables I, II, III, and IV. As seen, 2-hexene upon reacting with O_3 did not produce detectable aerosol. This agrees with the data of workers using open chain olefins in dark systems with O_3 and in photochemical systems with NO_2. Aerosols formed when O_3 reacted separately with 1,5-hexadiene, α-pinene, and cyclohexene. The cyclic compounds were also tested in the system, olefin $+ NO_2 + h\nu$, and produced aerosols. The systems, cyclohexanone $+ h\nu$ and cyclohexanone $+$ 1-hexene $+ h\nu$, did not produce detectable aerosols.

Since cyclic olefins and diolefins form light-scattering aerosols upon reacting with O_3 and open chain mono-olefins do not, this strongly indicates that a key step in the process is the generation of two active sites (*20*) on the same molecule. A subsequent step in the process seems to be copolymerizing of the diactive material with the original organic reactant. The kinetic data in Figures 1 and 2, which show α-pinene being used more rapidly than O_3, support this view.

In sunlight cyclohexanone did not produce aerosols while irradiating cyclohexene $+$ NO_2 did. Cyclohexanone $+ h\nu$ and cyclohexene $+$ O would presumably produce the same diradical,

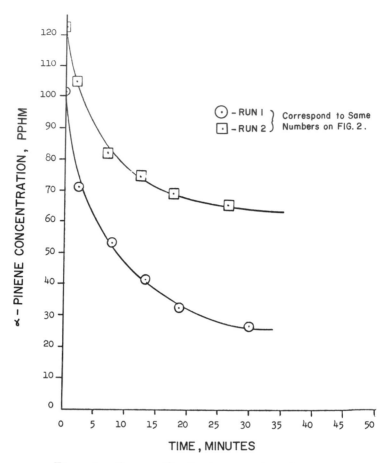

Figure 1.   Ozone utilization in ozone + α-pinene

$$
\begin{array}{ccccc}
& & & & & \text{O} \\
\text{H} & \text{H} & \text{H} & \text{H} & \text{H} & \| \\
\bullet\text{C}{-}\text{C}{-}\text{C}{-}\text{C}{-}\text{C}{-}\text{C}\bullet \\
\text{H} & \text{H} & \text{H} & \text{H} & \text{H}
\end{array}
$$

(*21, 22*), so that cyclohexene + O was not responsible for producing aerosol. The initial sequence for producing aerosol in the cyclohexene + NO_2 + $h\nu$ system therefore is:

$$NO_2 + h\nu \rightarrow NO + O \tag{1}$$

$$O + O_2 + M \rightarrow O_3 + M \tag{2}$$

$$O_3 + \text{cyclohexene} \rightarrow\rightarrow\rightarrow \text{aerosol} \tag{3}$$

Table IV.   Photochemical Systems in Air (RH = 0.1%)

Compound	Concentration, ppm	NO_2, ppm	Exposure Time, hr	Relative Light Scatter
Cyclohexanone	5	0.0	1	0
Toluene	5	0.1	1	0
1,5-Hexadiene	7	0.0	1	0
1,5-Hexadiene	7	0.5	1	1,000
Cyclohexanone + 1-Hexene	5 (both)	0.0	1	0

[a] Sunlight 1100–1500 hrs in North Carolina (February).

In the mid-latitudes the midday sun would produce several pphm O_3 from the radiation of 10 pphm of NO_2. We therefore agree with Groblicki (15) that even in the photochemical systems it is the O_3–olefin reactions and not the O–olefin reactions which produce aerosols from cyclic olefins.

The quantitative data in Tables II and III, regarding the effect of varying O_3 and cyclohexene, show that within limits producing aerosol depends directly on the concentration of both compounds.

Decreasing molecular oxygen (O_2) concentration in the reaction mixture from 20% to 2% decreases the particle formation (Table III). This indicates that incorporating O_2, perhaps as a peroxy radical, enhances the producing of aerosol. It might also indicate that O_3 was regenerated (*see* below).

Water vapor at the time of particle formation greatly affected the reaction (Table III). There was an increase in light-scattering, which visual microscopic examination suggested resulted from increased numbers of particles formed with increasing water vapor. In the humid runs particles were larger and coalesced more rapidly in the earlier reaction stages. This agrees with Goetz' finding (5) that the organic material

Table V.   Conditions and Rate Constants for Pinene–Ozone Reactions

Run Number	O_3 pphm	αPinene pphm	K[a] $pphm^{-1}\ hr^{-1}$	K[a] $liter$-$moles^{-1}\ sec^{-1}$
1	4.1	64.9	0.088	0.60 × 10⁵
2	4.0	105.8	0.202	1.37 × 10⁵
3	22.8	101.7	0.179	1.21 × 10⁵
4	26.5	122.1	0.129	0.88 × 10⁵
Average	—	—	0.15	0.99 × 10⁵

Run Number	O_3 pphm	βPinene pphm	K[a] $pphm^{-1}hr^{-1}$	K[a] $liter$-$moles^{-1}\ sec^{-1}$
5	2.4	94.9	0.058	3.94 × 10⁴

[a] Calculated from initial slope (runs 1, 2, and 3, O_3) (runs 4 and 5 hydrocarbon).

forms a film around the inorganic portion of the particle; the film resists coalescence. To orient the organic material into such a film takes a finite amount of time.

Kinetic information on the reaction of O_3 + α-pinene is given in Table V and Figures 1, 2, and 3. The value of the specific rate constant, using the data from the first few minutes of the reaction, is estimated to be 9.9×10^4 liters mole^{-1} sec^{-1} (0.15 pphm^{-1} hr^{-1}), assuming a second order reaction and a 1:1 stoichiometry.

Along with other workers using open chain olefins (23), we did not find an adherence to 1:1 stoichiometry. Earlier workers had interpreted this as interference in the O_3 measurements from peroxides and possibly other oxidants produced in olefin–O_3 reactions. However, in this study the deviation from 1:1 stoichiometry can not be interpreted in the same manner because the Regener-type chemiluminescence O_3 meter is considered specific for O_3, and we assumed that our readings were as accurate

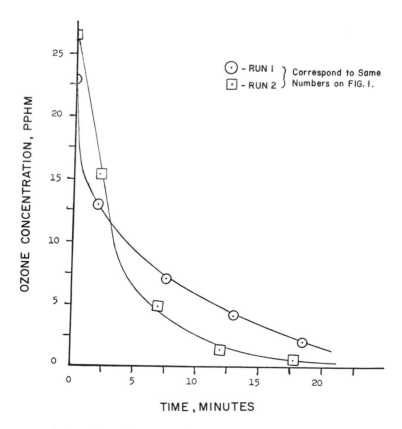

Figure 2.   α-Pinene utilization in ozone + α-pinene

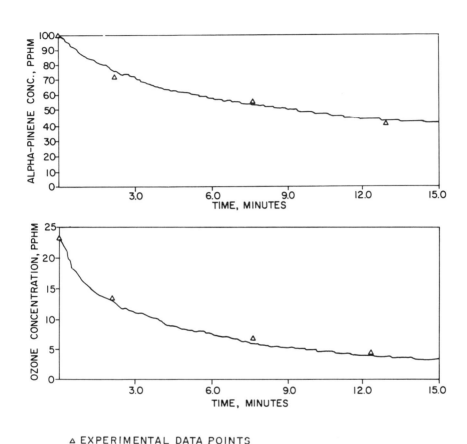

△ EXPERIMENTAL DATA POINTS
— COMPUTER PLOT
Figure 3.   Computer simulation in ozone + α-pinene

as the calibration. Also the ratio of α-pinene to O_3 consumed per unit time varied from time to time (from ~1 to ~50). If a 1:1 O_3 to α-pinene reaction is the first step in the process being observed, the empirical data suggest that the α-pinene reacted with a product of the α-pinene–O_3 reaction.

Numerous models were tested by computer simulation using numerical solution of a set of simultaneous differential equations and the one which best fitted the data was:

$$O_3 + \alpha p \rightarrow (\alpha p\text{—}O_x)^* \qquad (4)$$

$$\alpha p + (\alpha p\text{—}O_x)^* \rightarrow (\alpha p)_2 O_x \qquad (5)$$

$$(\alpha p\!-\!O_x)^* \rightarrow R \cdot + \text{ other products} \tag{6}$$

$$R \cdot + 2O_2 \rightarrow RO \cdot + O_3 \tag{7}$$

$r_1 = k \, (O_3) \, (\alpha p)$
$r_2 = k_2 \, (\alpha p\!-\!O_x)^* \, (\alpha p)$
$r_3 = k_3 \, (\alpha p\!-\!O_x)^*$
$r_4 = k_4 \, (R \cdot) \text{ (pseudo first order)}$
$k_1 = 0.3 \text{ pphm}^{-1} \text{ hr}^{-1} = 2.0 \times 10^5 \text{ liter mole}^{-1} \text{ sec}^{-1}$
$k_2 = 1.4 \text{ pphm}^{-1} \text{ hr}^{-1} = 9.5 \times 10^5 \text{ liter mole}^{-1} \text{ sec}^{-1}$
$k_3 = 96 \text{ hr}^{-1} = 1.6 \text{ sec}^{-1}$
$k_4 = 144 \text{ hr}^{-1} = 2.4 \text{ sec}^{-1} \text{ (pseudo first order)}$

A mechanistic generation of O_3 was needed because manipulation of the first models tested could not explain the behavior of the α-pinene and the O_3 simultaneously.

The rate constants were adjusted to force the computer simulation to fit the experimental data. Rate constant, k_1, was determined to be about 0.15 pphm^{-1} hr^{-1} ($\sim$10^5 liter mole^{-1} sec^{-1}) using the data from the first ten minutes of reaction. However, because of the rapidity of the subsequent reactions, the actual rate might be somewhat greater, and it was adjusted to twice the determined value in the final model. The other rate constants, k_2, k_3, and k_4 were originally estimated as 1.0 pphm^{-1} hr^{-1}, 48 hr^{-1}, and 96 hr^{-1}, respectively. Experience with the computer simulation caused us to revise the values upward to fit the theoretical curve to the data. The final values accepted were 1.4, 96, and 144. Figure 3 shows the computer plot and the experimental data.

This mechanism is suggested as a possible type of initiating series to generate aerosols from certain types of hydrocarbons. It is speculative but suggests reasons for the seemingly anomalous behavior of the reactants in the O_3–α-pinene reaction. The dimeric material in Step 2 above would be a precursor of the observed particles.

Although open chain mono-olefins do not produce light-scattering particles they apparently produce particles less than 0.1 in diameter. Both the system NO_2 + α-pinene + $h\nu$ (24) and photochemical systems containing NO_2 and open chain olefins—e.g. 1-hexene (25)—produced considerable quantities of condensation nuclei.

Figure 4 contains an infrared spectrum of the depositable aerosol formed in the O_3–α-pinene reaction. Figure 4 also contains an infrared spectrum of local rural ambient aerosol. There are large differences in the spectra; this agrees with Stephens (26) in comparing α-pinene aerosols with urban aerosols. Infrared spectra were obtained by Groblicki (15) for aerosol formed in the systems NO_2 + α-pinene + $h\nu$ and NO_2 + SO_2 + α-pinene. The spectra obtained in this laboratory for the O_3–α-pinene system and Groblicki's NO_2 + α-pinene + $h\nu$ system were nearly

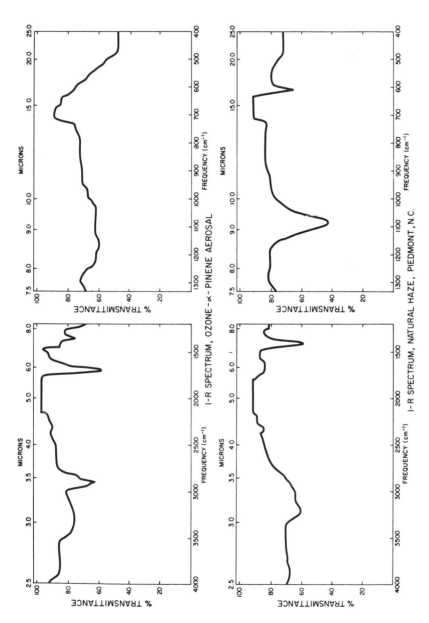

Figure 4.   Infrared spectra of depositable (MSI) aerosols

identical. This further supports Groblicki's suggestion that "the effects of reactant concentration and light intensity may be important only insofar as they affect the production of ozone" in the photochemical system without SO_2.

The similarity between Groblicki's photochemical aerosol from the system containing SO_2, NO, propylene, and (inadvertently) ammonia (NH_3) and these authors' ambient aerosol is striking. This similarity suggests that including SO_2 and possibly NH_3 in reaction mixture produces a synthetic aerosol which resembles the natural more, at least in IR spectra. Another series of experiments will be run to study possibly incorporating less reactive compounds into the aerocolloidal mass by reaction with the diactive species postulated above. This type of system could produce IR spectra quite different from a simple O_3–α-pinene aerosol since different functional groups are included.

Conclusions

The data reported here suggest that reactions between O_3 and cyclic olefins and O_3 and diolefins are important in incorporating organic vapors into the aerocolloidal mass. Within limits aerosol generation, as measured by light-scattering from deposited material, was enhanced by increasing concentrations of each reactant (O_3 and cyclohexene), O_2 (2–20%), and water vapor (0–100% RH, 21°C).

Comparing the behavior of open chained mono-olefins with cyclic and diolefins upon reaction with O_3 suggested that the following processes are operative in generating organic aerosols:

O_3 + cyclic olefin → diactive species
cyclic olefin + diactive species → dimeric material → aerosol.

Acknowledgment

This study was supported by funds from National Science Foundation Grant GA-14475. The authors wish to thank the Glidden Paint Company for furnishing the α-pinene utilized in this study.

Literature Cited

1. Goetz, A., Preining, O., "The Aerosol Spectrometer and Its Application to Nuclear Condensation Studies," pp. 164–182, Monograph 5, NASNRC No. 746, American Geophysical Union, Washington, D. C. (1960).
2. Cadle, R. D., "Atmospheric Chemistry of Chlorine and Sulfur Compounds," pp. 18–21, J. P. Lodge, Ed., Waverly, Baltimore, 1959.
3. Wayne, L. G., "The Chemistry of Urban Atmospheres," pp. 185–200, Los Angeles County Air Pollution Control District Technical Progress Report, Vol. III, Los Angeles, 1962.

4. Junge, C. E., "Air Chemistry and Radioactivity," pp. 178–180, Academic, New York, 1963.
5. Goetz, A., "Microphysical and Chemical Aspects of Sea Fog and Oceanic Haze," Final Report to U. S. Navy Weather Research Facility, Norfolk, 1966.
6. Leighton, P. A., "Photochemistry of Air Pollution," p. 174, Academic, New York, 1961.
7. Prager, M. J., Stephens, E. R., Scott, W. E., "Aerosol from Gaseous Air Pollutants," *Ind. Eng. Chem.* (1960) **52**, 521–524.
8. Rasmussen, R. A., Went, F. W., "Volatile Organic Material of Plant Origin in the Atmosphere," *Proc. N.A.S.* (1965) **53**, 215–220.
9. Coker, W. C., Totten, H. R., "Trees of the Southeastern States," pp. 16–33, University of North Carolina Press, Chapel Hill, 1945.
10. Mirov, N. T., "Composition of Gum Turpentines of Pines," U. S. Department of Agriculture, Forest Service Technical Bull. No. 1239, 1961.
11. Went, F. W., "Organic Matter in the Atmosphere and Its Possible Relationship to Petroleum Formation," *Proc. N.A.S.* (1960) **46**, pp. 212–221.
12. Jeffries, H. E., White, O., "α-Pinene and Ozone: Some Atmospheric Implications," Masters Thesis, University of North Carolina, Chapel Hill, 1967.
13. Ripperton, L. A., Worth, J. J. B., "Chemical and Environmental Factors Affecting Ozone in the Lower Troposphere," Final Report: National Science Foundation, Grant GA-1022 (1969).
14. Lillian, D., "The Effect of Water Vapor on the Photochemical System NO_2 + alpha-pinene + hv," Doctoral Dissertation, University of North Carolina, Chapel Hill, 1970.
15. Groblicki, P. J., Neble, G. J., "The Photochemical Formation of Aerosols in Urban Atmospheres," pp. 241–264, Charles S. Tuesday, Ed., American Elsevier, New York, 1971.
16. Altshuller, A. P., "Air Pollution," *Anal. Chem.* (1969) **41**, 1R–13R.
17. Mueller, P. K., Kothny, E. L., Pierce, L. B., Belsky, R., Imada, M., Moore, H., "Air Pollution," *Anal. Chem.* (1971) **43**, 1R–15R.
18. Robinson, E., Robbins, R. C., "Emissions Concentrations, and Fate of Particulate Atmospheric Pollutants," Stanford Research Project SCC-5507, 1971.
19. Regener, V. H., "Measurement of Atmospheric Ozone with the Chemiluminescent Method," *J. Geophys. Res.* (1964) **69**, 3795–3800.
20. Jaffee, S., Loudon, R., ADVAN. CHEM. SER. (1972) **113**, 264.
21. Calvet, J. G., Pitts, J. N., Jr., "Photochemistry," p. 409, Wiley, New York, 1966.
22. Wayne, L. G., Bryan, R. J., Weisburd, M., Danchick, R., "Comprehensive Technical Report on all Atmospheric Contaminants Associated with Photochemical Air Pollution," pp. 4-57–4-58, Systems Development Corporation, Santa Monica, 1970.
23. Leighton, P. A., *Ibid.*, p. 161.
24. Lillian, D., ADVAN. CHEM. SER. (1972) **113**, 211.
25. Decker, C. E., Page, W. W., "A Photochemical Study of Systems Containing Blends of Hexene-1, Nitrogen Dioxide and Sulfur Dioxide," Masters Thesis, University of North Carolina, Chapel Hill, 1965.
26. Stephens, E. R., Price, M. A., "Smog Aerosol: Infrared Spectra," *Science* (1970) **168**, 1584–1586.

RECEIVED May 24, 1971.

9

The Role of Carbon Monoxide in Polluted Atmospheres

M. C. DODGE and J. J. BUFALINI

Environmental Protection Agency, Research Triangle Park, N. C. 27711

The effect of CO on the rate of oxidation of NO has been investigated. Adding CO enhances the oxidation of NO in the presence and absence of reactive hydrocarbons. However, since CO concentrations required to achieve this conversion are higher than those usually present in polluted atmospheres, this work suggests that ambient levels of CO do not affect photochemical smog formation.

Until recently the role of carbon monoxide in polluted atmospheres has been largely ignored. Heicklen, Westberg, and Cohen (1) suggested that hydroxyl radicals, postulated as intermediates in the reaction of hydrocarbons with NO, serve as chain carriers to oxidize NO to NO_2 in the presence of CO:

$$OH + CO \rightarrow H + CO_2 \tag{1}$$

$$H + O_2 + M \rightarrow HO_2 + M \tag{2}$$

$$HO_2 + NO \rightarrow NO_2 + OH \tag{3}$$

This reaction sequence increases the rate of NO oxidation, thereby increasing the rate of photochemical smog formation and the ambient level of O_3 since less O_3 is consumed by reacting with NO. Westberg and Cohen (2) incorporated this reaction sequence in a computer program to estimate the effect of 100 ppm of CO on isobutene and NO_2. Their calculations predicted that the ozone concentration in this system is 50% greater than that in a similar system without CO.

The role of carbon monoxide in photochemical smog formation has been extensively studied by numerous investigators. Wilson and Ward (3), who studied the oxidation of nitric oxide in the presence of n-butane, n-butane and CO, and CO alone, found that adding 400 ppm of CO to 6.5 ppm n-butane and 0.6 ppm NO_x increased oxidant formation and decreased the time required for the NO_2 to reach a maximum. They also found that adding 400 ppm of CO to 0.5 ppm NO led to complete NO oxidation and ozone formation.

Westberg, Cohen, and Wilson (4) studied the CO effect on the photo-oxidation of isobutene. Adding 100 ppm CO to 3 ppm isobutene and 1.5 ppm NO decreased the time required to reach the maximum NO_2 concentration from 130 to 100 min. Ozone appeared much earlier when CO was present, but the peak ozone concentration was about the same as in the absence of carbon monoxide.

Stedman and coworkers (5) found that, even in the absence of hydrocarbons, CO greatly accelerated the conversion rate of NO to NO_2 when water was present in the system. When 2.3 ppm NO and 2500 ppm CO were irradiated in air of 25% relative humidity, these investigators found that the oxidation rate was about 7 times faster than in the absence of CO. This rapid speeding up in the NO oxidation rate was attributed to the forming of hydroxyl radicals via the reactions:

$$NO + NO_2 + H_2O \rightleftharpoons 2HONO \qquad (4)$$

$$HONO + h\nu \rightarrow OH + NO \qquad (5)$$

At least one study on the fate of CO in the atmosphere indicated that CO may affect the rate of NO oxidation very little. Dimitriades and Whisman (6) carried out irradiations of 1 ppm propylene, 1 ppm NO, and CO at 1, 10, and 100 ppm. Their results showed that CO did not noticeably affect the conversion rate of NO to NO_2.

Most evidence available before this investigation indicated that large concentrations of CO affect the rate of nitric oxide oxidation and perhaps the rate of hydrocarbon consumption and O_3 formation. Whether the role of carbon monoxide is important enough to affect photochemical smog formation is studied here.

Experimental

Samples used in this study were contained in 150-liter plastic bags made from fluorinated ethylene-propylene copolymer (Dupont FEP Teflon). The samples were prepared by injecting known amounts of the hydrocarbon, NO, and CO into a stream of purified air. Irradiations were carried out in one of two thermostatted irradiation chambers at $23 \pm 1°C$ and at ambient atmospheric pressure. One chamber was fitted with 26 GE F40BLB blacklights and the other contained 36 GE F42-T6 blacklights. The light intensities of the two irradiation chambers were determined by measuring the disappearance rate of NO_2 in a nitrogen atmosphere (7). The values obtained for k_d were 0.50 min^{-1} for one chamber and 0.35 min^{-1} for the other. One series of runs, carried out in a 335-ft^3 chamber, gave a k_d value of 0.40 min^{-1}.

All chemicals were chemically pure and were used without purifying further, except for carbon monoxide. CO was passed through packed columns to remove any $Fe(CO)_5$ impurity that might be in the tank since Westberg *et al.* (*4*) reported that the iron carbonyl impurity in CO enhanced the rate of nitric oxide oxidation. The purified CO used in this work showed no $Fe(CO)_5$ when run in a long path (10-meter) IR cell. J. T. Baker dry air was used in all runs.

All hydrocarbons were analyzed with a gas chromatograph equipped with a flame ionization detector. In those runs where carbon monoxide was monitored, the CO was passed over nickel in a stream of hydrogen to convert it to methane for detecting by flame ionization. Nitrogen oxides were measured by the Saltzman colorimetric procedure (*8*). Ozone formation was monitored with a chemiluminescence instrument containing a photomultiplier tube that detects the photons emitted when ozone reacts with ethylene (*9*). The water vapor content of the samples was analyzed with a Hygrodynamics, Inc. hygrometer, Model 15-3001, capable of measuring relative humidities from 5–100%. Unless otherwise stated, the relative humidity of all runs described in this paper was less than the detectable limit of the hygrometer.

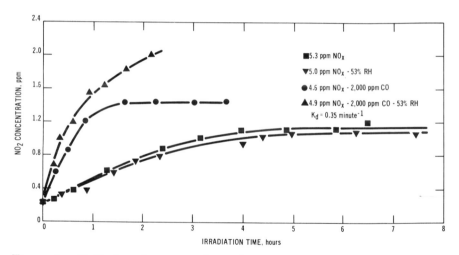

Figure 1. Oxidation of nitric oxide in the presence and absence of CO and water vapor

Results and Discussion

The effect of carbon monoxide on the oxidation of NO in the absence of hydrocarbons was studied. The formation of NO_2 in the presence and absence of CO and water vapor is shown in Figure 1. The k_d value for these four runs was 0.35 min⁻¹.

When NO was irradiated in the absence of CO and H_2O at an initial NO_x concentration of 5.3 ppm, NO_2 reached a constant value of 1.15 ppm after about 4 hours of irradiation. The presence of water vapor at 53%

relative humidity did not greatly affect the rate. However, the oxidation rate was greatly speeded up by the presence of 2,000 ppm CO, a puzzling result since CO should have no effect in a dry system where there are no hydroxyl radicals present. An explanation can be afforded by considering the various reactions occurring in this system. Those reactions involving only the oxides of nitrogen are:

$$NO_2 + h\nu \rightarrow NO + O \tag{6}$$

$$O + NO_2 \overset{M}{\rightarrow} NO_3 \tag{7}$$

$$O + NO_2 \rightarrow NO + O_2 \tag{8}$$

$$O + NO \overset{M}{\rightarrow} NO_2 \tag{9}$$

$$NO_3 + NO \rightarrow 2NO_2 \tag{10}$$

$$O + O_2 \overset{M}{\rightarrow} O_3 \tag{11}$$

$$O_3 + NO \rightarrow O_2 + NO_2 \tag{12}$$

$$NO_2 + O_3 \rightarrow NO_3 + O_2 \tag{13}$$

$$2NO + O_2 \rightarrow 2NO_2 \tag{14}$$

$$NO_3 + NO_2 \rightarrow N_2O_5 \tag{15}$$

$$N_2O_5 \rightarrow NO_3 + NO_2 \tag{16}$$

If H_2O is present in the system, the following reactions also occur:

$$NO + NO_2 + H_2O \underset{b}{\overset{a}{\rightleftharpoons}} 2HONO \tag{4}$$

$$HONO + h\nu \rightarrow OH + NO \tag{5}$$

$$N_2O_5 + H_2O \rightleftharpoons 2HNO_3 \tag{17}$$

$$OH + NO \overset{M}{\rightarrow} HONO \tag{18}$$

$$OH + NO_2 \overset{M}{\rightarrow} HNO_3 \tag{19}$$

By adding CO to the system, the following sequence occurs:

$$CO + OH \rightarrow CO_2 + H \tag{1}$$

$$H + O_2 \overset{M}{\rightarrow} HO_2 \tag{2}$$

$$HO_2 + NO \rightarrow NO_2 + OH \tag{3}$$

If the usual steady state approximations are made, the following expression for the rate of change of NO concentration with time is derived:

$$\frac{d(NO)}{dt} = -k_{14}(O_2)(NO)^2 + \frac{2k_8 k_a \phi (NO_2)^2}{k_{11}(O_2)(M)}$$

$$-k_{4a}(NO)(NO_2)(H_2O) + k_{4b}(HONO)^2 + k_5(HONO)$$

$$-\frac{k_5 k_{18}(NO)(HONO)}{k_{18}(NO)(M)+k_{19}(NO_2)(M)} - \frac{k_1 k_5(CO)(HONO)}{k_{18}(NO)(M)+k_{19}(NO_2)(M)}$$

At the beginning of the irradiation of NO and CO, even in a dry system, enough water vapor is present to make the last term of this equation the most important. Even if the permeation of water present in room air into the Teflon bag is excluded, tank air may contain up to 5 ppm water vapor, according to manufacturer's specifications. Assuming a steady state concentration of nitrous acid, 5 ppm water results in the forming of 3×10^{-3} ppm HONO. Using $k_1 = 8.9 \times 10^7$ l/mole-sec (*10*) and using the values assumed by Westberg (*11*) ($k_5 = 0.1$ $k_a \phi = 4.0 \times 10^{-4}$ sec^{-1}, $k_{18} = 1 \times 10^9$ l^2/mole^2-sec, and $k_{19} = 3 \times 10^9$ l^2/mole^2-sec), this term becomes 0.06 ppm/min for a CO concentration of 2,000 ppm. The observed value for the oxidation rate of NO in the presence of 2,000 ppm is 0.03 ppm/min. Despite the many assumptions involved in this calculation, we believe the result shows that the rapid acceleration in the NO oxidation rate is explained by the presence of a trace amount of water vapor impurity.

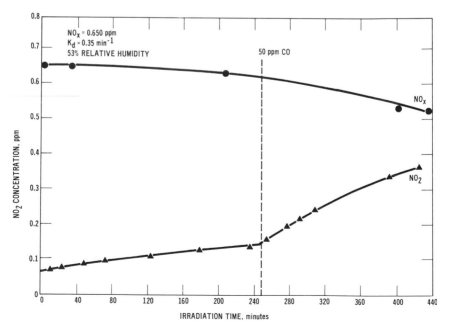

Figure 2.   Effect of adding 50 ppm CO to 0.650 ppm NO$_x$

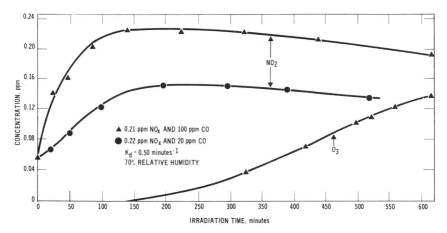

Figure 3.   Irradiation of 0.2 ppm NO_x with 20 and 100 ppm CO at 70% relative humidity

Figure 1 also shows the growth of NO_2 in a system containing 2,000 ppm CO at 53% relative humidity. As expected, the rate of formation of NO_2 was greatly enhanced over that of the dry system.

The effect of adding CO to nitric oxide in air of 53% humidity at concentrations more nearly those encountered in polluted atmospheres is shown in Figure 2. Adding 50 ppm CO to the system after 4 hours of irradiating rapidly accelerated the NO oxidation rate.

Figure 3 presents data for the oxidations of 0.2 ppm of NO_x with 20 ppm CO and 100 ppm CO at 70% humidity and a k_d value of 0.50 min⁻¹. As indicated, when 100 ppm CO was present, NO was completely oxidized to NO_2, and O_3 was produced. Wilson and Ward (3) also detected the formation of ozone in a similar system. In the run containing 20 ppm CO (the concentration usually found in early morning automobile traffic), the oxidation rate of NO was much slower than that observed with 100 ppm CO, and no ozone was produced. Nevertheless, NO oxidation in this run occurred much faster than NO oxidation in the absence of CO.

This study shows that, in the absence of hydrocarbons, ambient levels of carbon monoxide enhance the oxidation rate of nitric oxide.

The effect of CO on hydrocarbon and NO_x systems was extensively studied. The hydrocarbons studied ranged from the more unreactive paraffins and aromatics to the very reactive unsaturated hydrocarbons. Usually the hydrocarbon–NO_x ratios (HC as total C) were representative of those found in polluted atmospheres although the absolute concentrations of the substances were much higher than ambient levels.

All of the hydrocarbon–NO_x systems investigated here were irradiated with light of intensity, $k_d = 0.50$ min⁻¹.

Mesitylene. The data for the mesitylene–NO_x system are shown in Table I. Adding CO did not significantly affect the time required for NO_2 to reach a maximum or the time required for one-third of the mesitylene to react, as shown in the fourth and fifth columns. In contrast to theory, however, adding 2,000 ppm CO seemed to decrease the amount of ozone formed. This decrease may be a function of the mesitylene–NO_x ratio rather than a function of the CO concentration. M/NO_x in this run was 1.8, but in the other three runs it was 2.1. As the hydrocarbon–NO_x ratio decreases, the amount of ozone formed also decreases (12).

Table I.   Photooxidation of Mesitylene–NO_x–CO Mixtures

Initial Concentrations			$\dfrac{Mesitylene}{NO_x}$	NO_2 t max, min.	Mesitylene t 1/3, min.	Max O_3, ppm
Mesitylene, ppm	NO_x, ppm	CO, ppm				
9.62	4.58	0	2.1	19	27	1.1
8.80	4.18	75	2.1	19	29	1.0
9.20	4.45	600	2.1	18	30	0.95
9.10	5.00	2000	1.8	17	32	0.80

1-Butene. The effect of adding 2,000 ppm CO to 9 ppm 1-butene and 3.5 ppm NO_x is shown in Figure 4. The time required for NO_2 to maximize decreased from 45 to 22 min. The half-life of the 1-butene decreased from 60 to 41 min. Also, by adding 2,000 ppm CO, the peak ozone concentration increased from 1.15 to 1.40 ppm.

The complete data on the 1-butene–NO_x system (*see* Table II) are scattered because of the varying 1-butene–NO ratios in these runs, but certain trends are apparent. Adding CO decreased the time required to reach the NO_2 maximum concentration and the half-life of the 1-butene. Also the amount of O_3 produced seemed to increase with increasing CO concentration. The data also show that adding water vapor decreased the peak ozone concentration; this result agrees with the results of Wilson and Levy (13) who also observed that less ozone was produced by irradiating 1-butene with NO_x at higher relative humidities.

Ethylene. CO affected the photooxidation of ethylene in the presence of oxides of nitrogen similarly to the 1-butene system. When 10.8 ppm ethylene and 2.8 ppm NO_x were irradiated in the absence of CO, the maximum NO_2 concentration occurred at approximately 50 min., and the ethylene half-life was 130 min. Ozone began to appear when the NO_2 was maximum and increased to a peak value of 1.40 ppm at 160 min. When the same concentrations of ethylene and NO_x were irradiated in the presence of 500 ppm CO, the time for the NO_2 to maximize was shortened to 35 min., the ethylene half-life decreased to 105 min., and the peak ozone concentration increased to 1.60 ppm.

The rate parameters for the ethylene–NO_x system as a function of CO concentration are presented in Figure 5. The time required to reach the maximum NO_2 concentration remained rather constant up to about 100 ppm CO and then changed rapidly up to 1,000 ppm. Adding more CO did not greatly change the time required to reach the maximum; the same was true of the time required to reach the peak O_3 concentration. The ethylene half-life depended only slightly upon the CO concentration. The time required for one-half of the ethylene to react decreased from about 130 to 105 min. by adding 500 ppm CO; adding more had no effect. The data obtained for the peak ozone concentration are scattered, but it seems generally that, overall, the ozone concentration was unaffected by the presence of CO.

Toluene. The effect of CO on the less reactive toluene–NO_x system was also studied. Toluene at a concentration of 10 ppm and NO_x at 3.5 ppm were irradiated with 0–1,000 ppm CO. The time required for NO_2 to maximize decreased from 180 min. when no CO was present to 160 min. when 500 ppm CO was added to the system and the O_3 peak concentration increased from 0.80 to 1.10 ppm. When only 100 ppm CO was added, no effect was observed.

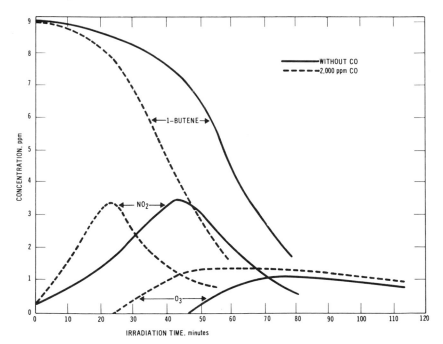

Figure 4.   Irradiation of 1-butene and NO in the presence and absence of CO

Table II.   Photooxidation of

	Initial Concentrations			
1-Butene, *ppm*	NO_x, *ppm*	NO_2, *ppm*	*CO,* *ppm*	*Relative humidity,* *%*
9.7	3.90	0.70	0	0
9.0	3.45	0.35	0	0
9.0	3.75	0.30	0	0
9.0	3.00	0.50	0	70
9.2	3.50	0.45	500	0
9.0	3.65	0.63	1,000	0
9.0	3.20	0.42	1,000	70
9.0	3.65	0.65	2,000	0
9.4	3.80	0.45	2,000	0
8.8	3.50	0.35	2,000	70
8.8	2.85	0.40	2,000	70

Benzene. The unreactive benzene–NO_x system was also studied in the presence and absence of CO. When 5 ppm benzene and 4.5 ppm NO_x were irradiated in the absence of CO, only about 1% of the benzene reacted in seven hours. The curve for the growth of NO_2 was the same as that shown in Figure 1 for irradiating NO in the absence of hydrocarbons. When benzene and NO were irradiated with 2,000 ppm CO, about 5% of the benzene reacted after 7 hours, and there was some increase in the oxidation rate of NO. However the NO_2 growth curve was almost identical to the one shown in Figure 1 for irradiating NO_x in the presence of 2,000 ppm CO. These results suggest that benzene affected the oxidation rate of NO very little.

Pentane. The systems n-pentane–NO_x and 2,3,4-trimethylpentane–NO_x were studied in the presence and absence of carbon monoxide. In both systems 2,000 ppm CO added to 10 ppm paraffin and 3 ppm NO_x increased the oxidation rate of NO but did not affect the reactivity of the paraffin. The increase in the oxidation rate of NO was the same as that shown in Figure 1 for 2,000 ppm CO and 4.6 ppm NO_x in the absence of hydrocarbon; ozone was not formed in either system.

Hydrocarbon–NO_x Systems Approximating Ambient Polluted Conditions

All hydrocarbon–NO_x systems presented thus far were studied at concentrations almost ten times greater than the levels found in polluted urban air. In each case more than 200 ppm CO (approximately ten times the ambient level of 20 ppm found in polluted areas) was required before

1-Butene–NO$_x$–CO Mixtures

$\dfrac{1\text{-}Butene}{NO}$	NO_2 t max, min.	Max O_3, ppm	1-Butene t ½, min.
3.0	37	1.30	55
2.9	45	1.15	60
2.6	51	1.30	67
3.6	45	1.05	61
3.0	36	1.40	51
3.0	32	1.45	50
3.2	32	1.10	51
3.0	22	1.40	41
2.8	25	1.40	41
2.8	31	1.05	52
3.6	28	1.00	50

a noticeable effect occurred. It was of interest to see if, by working at nearly ambient concentrations of hydrocarbon and NO$_x$, smaller concentrations of CO approximating those of morning traffic conditions could cause a noticeable effect. For this reason, 1 ppm ethylene and 0.2 ppm NO$_x$ were irradiated with 0–100 ppm CO. The effect of adding 50 ppm CO is shown in Figure 6. The time required for NO$_2$ to reach a maximum decreased from 100 to approximately 80 min. and the peak ozone concentration increased from 0.56 ppm to 0.68 ppm. Although a CO effect was observed in this run, the reactivity of nitric oxide or the level of oxidant formation was not affected in runs carried out in the presence of only 10–20 ppm CO, concentrations more nearly approximately ambient levels.

Experiments conducted in the large 335-ft^3 irradiation chamber also showed that in most cases the effect of ambient concentrations of CO on the oxidation rate of NO to NO$_2$ is negligible. The results for three series of runs, one involving only paraffins, one involving paraffins and more reactive hydrocarbons, and one involving no hydrocarbons, are shown in Table III. The various hydrocarbons and their relative concentrations were chosen to represent the Los Angeles atmosphere as determined by Kopczynski and co-workers (*14*). All runs were conducted at 50% relative humidity.

The data shown in Table III for the series of runs involving only parafins are scattered. For the run containing only 20 ppm CO, there seemed to be a negative effect on the oxidation rate of NO and the oxidant yield. The time to oxidize one-half of the NO seemed to increase, and the amount of oxidant formed seemed to decrease. However, these numbers probably are within experimental error of being the same. A large positive effect occurred for the run with 100 ppm CO. The NO oxidation rate greatly increased and the amount of oxidant formed doubled. However, though an effect did occur, 100 ppm CO is too high to represent the ambient polluted atmosphere. Even during severe Los Angeles smog episodes the concentration of CO seldom exceeds 20 ppm.

For the series of runs containing paraffins, olefins, and aromatics, less of a CO effect was found. Again, for the run containing only 20 ppm CO, no effect was observed. When 100 ppm CO was present, the time required to oxidize one-half of the NO decreased from 37 to 26 min., and the maximum concentration of oxidant formed increased from 54 to 66 pphm. This effect is much less than that observed for the paraffin series.

For the third series of runs shown in Table III, where NO_x was irradiated in the absence of hydrocarbons, there was a large CO effect. The time to oxidize one-half of the NO decreased from greater than 300 min. when no CO was present to 186 min. when 20 ppm CO were added. When 50 or 100 ppm CO were present, all of the NO was oxidized to NO_2 in the run, and the formation of ozone occurred.

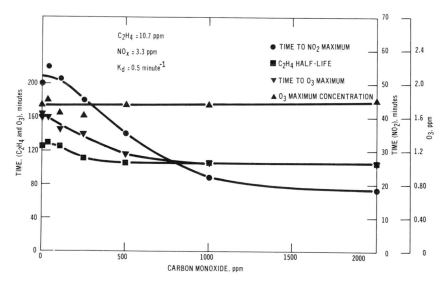

Figure 5. Photooxidation of ethylene–NO_x systems as a function of CO concentration

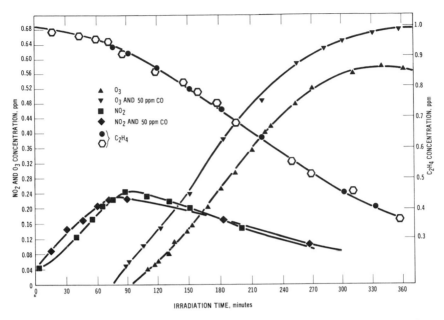

Figure 6. *Irradiation of 1 ppm ethylene and 0.2 ppm NO$_x$ with 0 and 50 ppm CO*

Table III. Photooxidation of Hydrocarbon–NO$_x$–CO Mixtures in Large Irradiation Chamber

Initial Concentrations			*Time to Oxidize* $\frac{1}{2}$ *NO$_o$, min.*	*Max Oxidant (Corrected for NO$_2$), pphm*	*5 Hr Oxidant Dosage ppm-min.*
NO$_x$, pphm	*CO, ppm*	*Hydrocarbons, ppm C*			
52	0	10 paraffins	85	13	11
50	20	10 paraffins	96	11	9
54	100	10 paraffins	51	24	27
50	0	2¼ paraffins ¾ olefins 2 aromatics	37	54	77
50	20	2¼ paraffins ¾ olefins 2 aromatics	36	55	89
54	100	2¼ paraffins ¾ olefins 2 aromatics	26	66	119
55	0	0	> >300	0	0
55	20	0	186	0	0
57	50	0	116	4	2
55	100	0	65	21	19

In this series of runs, the less reactive the system, the greater the effect of CO on the rate of oxidation of NO and the level of oxidant formation.

Conclusions

The results of these studies indicate that, in the absence of hydrocarbons, ambient levels of carbon monoxide speed up the rate of nitric oxide oxidation. However, when hydrocarbons are present, ambient concentrations of CO do not noticeably increase oxidant formation or NO oxidation rate. This result is explained by considering the two reactions of importance in this system:

$$CO + OH \rightarrow CO_2 + H \qquad (1)$$

$$HC + OH \rightarrow Products \qquad (20)$$

Whether or not CO can affect an increase in the oxidation rate of NO in the presence of hydrocarbons depends on the relative rates of these competing reactions. For a highly reactive hydrocarbon such as mesitylene, the reaction of the hydrocarbon with hydroxyl radicals is so fast that the reaction of CO with OH cannot compete even at high CO–hydrocarbon ratios. For less reactive hydrocarbons such as ethylene and 1-butene, CO competes with the hydrocarbon for the OH radicals and, in systems containing these hydrocarbons, a carbon monoxide effect is possible. The rate constant for the reaction of ethylene with hydroxyl radicals has been measured to be 3.6×10^9 l/mole-sec (15). This is forty times greater than the rate constant of 8.9×10^7 l/mole-sec (10) for the reaction of OH with CO. Therefore, a CO effect should be possible at CO–ethylene ratios of 40 or greater. Experimentally, an increase in the NO oxidation rate for this system was observed at a CO–hydrocarbon ratio of 50.

For unreactive hydrocarbons such as benzene and the pentanes, the reaction between hydroxyl radicals and the hydrocarbon is so slow that most of the hydroxyl radicals react with CO rather than with the hydrocarbon. In these systems the rise in the NO oxidation rate is the same as that observed in the absence of the hydrocarbon.

In summary the reaction of CO with OH radicals does not seem to be rapid enough to compete with the reactions of hydrocarbons with OH radicals in the polluted atmosphere.

Acknowledgment

The authors are grateful to Mr. Stanley L. Kopczynski for the use of his data obtained in the large irradiation chamber.

Literature Cited

1. Heicklen, J., Westberg, K., Cohen, N., "The Conversion of NO to NO₂ in Polluted Atmospheres," *Symp. Chem. Reactions Urban Atmospheres,* Research Laboratories, General Motors, Warren (October 6–7, 1969).
2. Westberg, K., Cohen, N., "The Chemical Kinetics of Photochemical Smog as Analyzed by Computer," The Aerospace Corp., El Segundo (December 1969).
3. Wilson, W. E., Jr., Ward, G. F., "The Role of Carbon Monoxide in Photochemical Smog," *Proc. 160th Meet., Amer. Chem. Soc.,* Chicago (September 13–18, 1970).
4. Westberg, K., Cohen, N., Wilson, K. W., *Science* (1971) **171**, 1013.
5. Stedman, D. H., Morris, E. D., Jr., Daby, E. E., Nicki, H., Weinstock, B., "The Role of OH Radicals in Photochemical Smog Reactions," *Proc. 160th Meet., Amer. Chem. Soc.,* Chicago (September 13–18, 1970).
6. Dimitriades, B., Whisman, M., *Environ. Sci. Technol.* (1971) **5**, 219.
7. Tuesday, C. S., Cadle, R. D., Eds., "Chemical Reactions in the Lower and Upper Atmospheres," pp. 1–49, Interscience, New York, 1961.
8. Saltzman, B. E., *Anal. Chem.* (1954) **26**, 1949.
9. Nederbragt, G. W., Van der Horst, A., Van Duijn, J., *Nature* (1965) **206**, 87.
10. Greiner, N. R., *J. Chem. Phys.* (1967) **46**, 2795.
11. Westberg, K., private communication, 1971.
12. Glasson, W. A., Tuesday, C. S., *Environ. Sci. Technol.* (1970) **4**, 37.
13. Wilson, W. E. Jr., Levy, A., *J. Air Pollution Control Assoc.* (1970) **20**, 385.
14. Kopczynski, S. L., Lonneman, W. A., Sutterfield, F. D., Darley, P. E., *Environ. Sci. Technol.,* to be published.
15. Greiner, N. R., *J. Chem. Phys.* (1970) **53**, 1284.

RECEIVED July 23, 1971. Mention of company or commercial products does not constitute endorsement by the Environmental Protection Agency.

10

The Chemiluminescent Reactions of Ozone with Olefins and Organic Sulfides

J. N. PITTS, JR., W. A. KUMMER,[a] R. P. STEER,[b] and B. J. FINLAYSON

University of California, Riverside, Calif. 92502

Spectra from the chemiluminescent gas phase reactions at 0.5 torr, of ozone with ethylene, tetramethylethylene, trans-*2-butene, and methyl mercaptan at room temperature are presented, and a summary of the general features of the emissions obtained from reaction in the gas phase of ozone with fourteen different olefins is given. The emitting species in the ozone–olefin reactions have been tentatively identified as electronically excited aldehydes, ketones, and α-dicarbonyl compounds. The reaction of ozone with hydrogen sulfide, methyl mercaptan, and dimethylsulfide produces sulfur dioxide in its singlet excited state.*

The increasing severity of urban air pollution has recently led to the development of new methods for the sensitive and specific measurement of the low concentrations of ozone found in urban atmospheres. The most important methods monitor ozone by detecting the chemiluminescence produced in the reaction of ozone with some organic substrate. Regener's method (*1, 2*) monitors the intensity of the chemiluminescence produced by ozone reacting with rhodamine-β adsorbed on a silica gel disc. The method of Fontijn *et al.* (*3*) follows the intensity of emission from the reaction of ozone with nitric oxide at low pressures. The most convenient method (*4*) seems to be that developed by Nederbragt *et al.* (*5*) and Warren and Babcock (*6*) where the emission intensity from the chemiluminescent reaction of ozone and ethylene at atmospheric pressure is monitored. All of these techniques have been evaluated recently by Hodgeson *et al.* (*4*).

[a] Present address: SIBA-Geigy, Photochemi LTD, Fribourg, Switzerland.
[b] Present address: Department of Chemistry and Chemical Engineering, University of Saskatchewan, Saskatoon, Saskatchewan, Canada.

Although higher olefins do not produce detectable chemiluminescence when reacting with ozone at atmospheric pressure (7) at pressures of about one-half torr, light is emitted by these reactions (8). Several organic sulfides also give an emission when they react with ozone at these pressures.

These reactions may be important for several reasons. It may be possible to use the chemiluminescent reaction of ozone with organic sulfides to monitor the low concentrations of sulfur compounds in urban atmospheres. Also, excited species are being formed, and these reactive intermediates may be important in high altitude atmospheric reactions. Finally, identifying these emitting species should give information about the mechanisms of gas phase ozone reactions. Current progress on these reactions by the authors is reviewed here.

FLOW SYSTEM AND DETECTION APPARATUS FOR CHEMILUMINESCENCE STUDIES

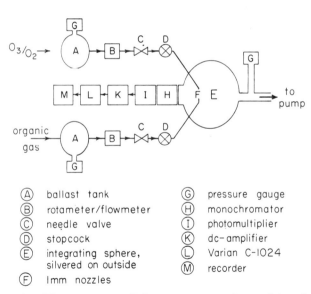

(A) ballast tank
(B) rotameter/flowmeter
(C) needle valve
(D) stopcock
(E) integrating sphere, silvered on outside
(F) 1mm nozzles
(G) pressure gauge
(H) monochromator
(I) photomultiplier
(K) dc-amplifier
(L) Varian C-1024
(M) recorder

Figure 1. Flow system and detection system for studying the chemiluminescent reactions of ozone with olefins and organic sulfides

Experimental

The flow system used in these studies is shown in Figure 1. Approximately 2% (V/V) of ozone in oxygen was produced by passing oxygen (Matheson, ultra-high purity grade) through a Welsbach ozone generator. It was then stored until needed in a five liter storage bulb. Dur-

Table I. Summary of Visible Emission Spectra Obtained from the Reaction of Ozone with Olefins

Type	Olefin	Emission Spectrum Characteristics
A		Broad; peak at approximately 440 nm
B		Narrow; peak at 520 nm with broad shoulders at 465 nm and 565 nm
C		Narrow; peak at 520 nm, with smaller peaks at approximately 565 nm and 595 nm. Overlapped by broad peak at 465 nm

ing a run the ozone–oxygen mixture flowed continuously into the reaction vessel through a 1 mm nozzle. The flow rate was controlled by a flow-meter–needle valve–stopcock combination.

Trans-2-butene (J. T. Baker, 99.0%), ethylene (Matheson, 99.98% mole typical lot purity), and methyl mercaptan (Matheson, purity 99.5% minimum) were used as received. Tetramethylethylene (Chemical Samples Co., 99% purity) was purified by passing through an alumina column to remove contaminating oxidation products. All olefins were degassed and stored in a five liter bulb. The sulfur compounds were stored without degassing. The flow rates of the organic substrates were also controlled by a flowmeter–needle valve–stopcock combination. During a run the organic substrate flowed continuously into the reaction vessel through a 1 mm nozzle.

The reaction vessel consisted of a 3 liter borosilicate glass flask which was silvered on the outside for increased light collecting efficiency. Light emission in the reaction vessel was observed through a 5.1 cm diameter planar borosilicate glass window. Light passed from the reaction flask into a 0.3 meter McPherson scanning monochromator–photomultiplier (EMI 9656KA) combination. The photomultiplier output was fed to a D.C. amplifier circuit with variable time constants, and the amplified output was displayed on a potentiometric recorder. At these low pressures (~0.5 torr) the emissions were generally so weak that the use of a

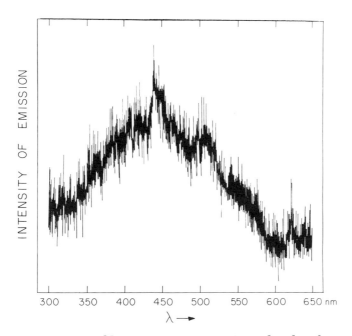

Figure 2. Visible emission spectrum from the chemiluminescent reaction of ozone with ethylene at room temperature (uncorrected for spectral sensitivity). Total pressure 0.4 torr; flow rate of O_3/O_2 is 30 cc/min; flow rate of ethylene 5 cc/min; spectral slit width 10.6 nm.

time-averaging computer was necessary to extract the emission spectrum from the background noise.

Contamination of the system, particularly after using organic sulfides, necessitated thorough cleaning of the system with organic solvents and aqueous hydrofluoric acid after each run. The total pressure in the reaction vessel varied between 0.2–0.8 torr. Typical flow rates ranged from 5–7 cc min^{-1} for the organic substrate and 20–65 cc min^{-1} for the ozonized oxygen. All experiments were performed at room temperature.

Results and Discussions

Table I summarizes the primary features of the chemiluminescent emission spectra obtained from the reaction of ozone with 14 simple olefins. The observed spectra fall into three classes which correlate somewhat with the olefin structure. Class A in Table I includes the three terminal olefins studied; all gave a broad, weak emission, peaking at about 440 nm. Figure 2 shows the spectrum obtained in the reaction of ozone with ethylene, a typical member of class A, at a total pressure of 0.4 torr. The emission spectrum may result from excited formaldehyde [emission

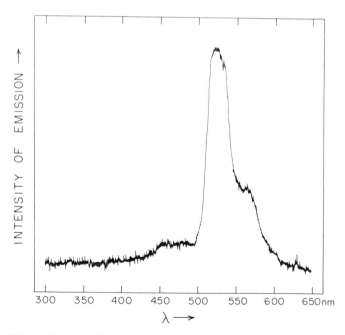

Figure 3.   Visible emission spectrum from the chemiluminescent reaction of ozone with tetramethylethylene at room temperature (uncorrected for spectral sensitivity). Total pressure 0.8 torr; flow rate of O_3/O_2 is 60 cc/min; flow rate of tetramethylene 5 cc/min; spectral slit width 5.3 nm.

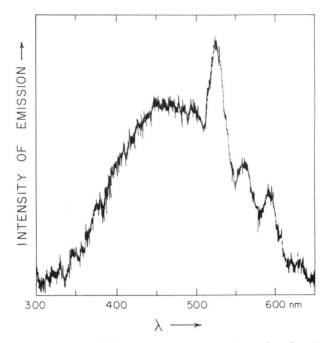

Figure 4. Visible emission spectrum from the chemi-luminescent reaction of ozone with trans-2-butene *at room temperature (uncorrected for spectral sensitivity). Total pressure 0.4 torr; flow rate of O_3/O_2 is 65 cc/min; flow rate of* trans-2-butene *5 cc/min; spectral slit width 10.6 nm.*

maximum about 424 nm (9)] or excited glyoxal [emission maximum about 478 nm (10)] or possibly from both.

The emission spectra produced by each reaction in Class A could not be compared in detail because of the low signal to noise ratio obtained in the experiment and the broad structure of the observed emission. They will be compared using more detailed spectra which are now being recorded.

Class B in Table I includes seven olefins, which are characterized by dialkyl substitution at one of the carbons of the olefinic double bond. The emission spectrum produced by reaction of these olefins with ozone is characterized by a narrow band peaking at about 520 nm with broad shoulders at 465 nm and 565 nm. Figure 3 gives the chemiluminescent emission spectrum obtained by reaction of a typical member of Class B, tetramethylethylene, with ozone at a pressure of 0.8 torr. It is similar to the fluorescent emission spectrum of biacetyl [broad emission from 440–495 nm (11)] combined with the phosphorescent emission of the same compound [narrow peaks at about 512 and 561 nm and a broad

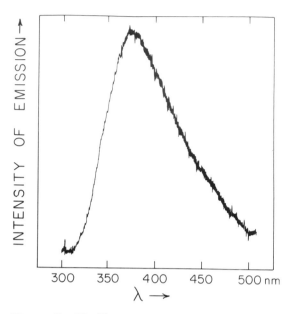

Figure 5. Visible emission spectrum from the chemiluminescent reaction of ozone with methyl mercaptan at room temperature (uncorrected for spectral sensitivity). Total pressure 0.2 torr; flow rate of O_3/O_2 is 20 cc/min; flow rate of methyl mercaptan 7 cc/min; spectral slit width 1.31 nm.

shoulder at 607 nm (12)]. Unpublished kinetic data from this laboratory and radiative lifetimes from the literature suggest that biacetyl is the observed intermediate. The result is equivalent to the addition of a molecule of oxygen across the double bond and the loss of two alkyl groups, an unusual ozonolysis reaction. These identifications of the emitting intermediates are tentative; further work is need to verify them.

Class *C* of Table I includes four olefins which, upon reacting with ozone, gave emission spectra characterized by a broad band peaking at 465 nm, which overlaps a narrower band centered at 520 nm. Smaller peaks occur at 565 and 595 nm. Figure 4 gives the spectrum produced by ozone reacting with a member of Class *C*, *trans*-2-butene, at a pressure of 0.4 torr. The three narrow peaks at the higher wavelengths are similar to the phosphorescent emission of biacetyl (12) while the broad peak at 465 nm resembles the emission from acetaldehyde [broad emission peaking at approximately 420 nm (13)]. Again there seems to be a cleavage of the olefinic double bond to produce excited acetaldehyde and an addition of oxygen across the double bond of the olefin to produce excited biacetyl.

Although α-dicarbonyl compounds are not known to be products of the ozonolysis of olefins, biacetyl has been isolated in photochemically initiated reactions (*14, 15*) which result in the net oxidation of olefins in the gas phase. For example, when a mixture of *cis*-2-butene, nitric oxide, and air is irradiated, small amounts of biacetyl are isolated. One of the pathways suggested to explain the production of biacetyl involves the reaction of ozone with *cis*-2-butene (*14*):

$$NO_2 \xrightarrow{h\nu} NO + O \tag{1}$$

$$O + O_2 \xrightarrow{M} O_3 \tag{2}$$

$$\diagup\!\!\!\!\diagdown + O_3 \longrightarrow \underset{\underset{H}{|}}{\overset{CH_3}{\overset{|}{C}}}\diagdown \underset{O}{} \diagup \underset{\underset{H}{|}}{\overset{O-O}{\overset{|}{C}}}\overset{CH_3}{} \longrightarrow \overset{CH_3}{\underset{\underset{O}{||}}{\overset{|}{C}}} - \overset{CH_3}{\underset{\underset{O}{||}}{\overset{|}{C}}} + H_2O \tag{3}$$

The chemiluminescence spectrum obtained from the reaction of ozone with methyl mercaptan at a pressure of 0.2 torr is shown in Figure 5. Reaction of hydrogen sulfide with dimethylsulfide with ozone give identical spectra consisting of a broad structureless band centered at approximately 370 nm (uncorrected for spectral sensitivity of the detection system). We have recently shown that this emission is identical to the fluorescence spectrum of sulfur dioxide (*16*). Since ozone oxidizes hydrogen sulfide to sulfur dioxide and water in the gas phase (*17, 18*), this result is not surprising.

As a result of the longer lifetimes of triplet states of electronically excited organic molecules as compared with their lowest excited singlet

Table II. Relative Emission Intensities in the Chemiluminescent Reactions of Ozone with Some Organic Compounds (*8*)

Reactant	Relative Integrated Emission Intensity[a]
ethylene	1
trimethylethylene	50
tetramethylethylene	50
cis- or *trans*-butene-2	10
2,3-dimethylbutadiene	8
2,5-dimethly-2,4-hexadiene	30
hydrogen sulfide	25
dimethyl sulfide	200
methyl mercaptan	2000
2,5-dimethylfuran	40

[a] Relative to ethylene = 1.

states, the phosphorescent emission produced in the reaction of the higher olefins is expected to be quenched at atmospheric pressure. For the ethylene and possibly the sulfide reactions, however, fluorescent emission from the short-lived singlet states may predominate over quenching processes even at atmospheric pressure. In the latter case it is then possible to operate a chemiluminescent detector at atmospheric pressure.

Kummer *et al.* (8) have reported that at pressures of about 0.5 torr, the relative emission intensities of the higher olefins and of the organic sulfides were substantially greater than that of ethylene; Table II summarizes the reported relative emission intensities. Since a recently developed commercial ozone monitor is based on the chemiluminescent reaction between ozone and ethylene, this suggests the possibility of using the sulfide–ozone chemiluminescent reaction to monitor the low concentration of sulfur compounds in ambient air. This possibility is being further investigated now.

Acknowledgment

This work was supported by the U. S. Department of Defense Grant Themis N00014-69-A-0200-500 and Grant AP00109, Research Grants Branch, Air Pollution Control Office, Environmental Protection Agency.

Literature Cited

1. Regener, V. H., *J. Geophys. Res.* (1960) **65**, 3975.
2. *Ibid.* (1964) **69**, 3795.
3. Fontijn, A., Sabadell, A. J., Ronco, R. J., *Anal. Chem.* (1970) **42**, 575.
4. Hodgeson, J. A., Martin, B. E., Baumgardner, R. E., *Proc.* 160th Meet. Amer. Chem. Soc., Chicago, 1970.
5. Nederbragt, G. W., Van der Horst, A., Van Duijn, J., *Nature* (1965) **206**, 87.
6. Warren, G. J., Babcock, G., *Rev. Sci. Instru.* (1970) **41**, 280.
7. Hodgeson, J. A., Air Pollution Control Office, private communication, 1971.
8. Kummer, W. A., Pitts, Jr., J. N., Steer, R. P., *Environ. Sci. Technol.* (1971) **5**, 1045.
9. Shuvalov, V. F., Vasilev, R. F., Postnilev, L. M., Shlapintokl, V. Y., *Dokl. Akad. Nauk. SSSR* (1963) **148** (2), 388.
10. Thompson, H. W., *Trans. Faraday Soc.* (1940) **36**, 988.
11. Longin, P., *C. R. Acad. Sci. Paris* (1968) **267B**, 128.
12. Longin, P., *C. R. Acad. Sci. Paris* (1968) **267B**, 404.
13. Longin, P., *C. R. Acad. Sci.* (1960) **251**, 2499.
14. Stephens, E. R., Statewide Air Pollution Research Center, private communication, 1971.
15. Altshuller, A. P., Cohen, I. R., *Intern. J. Air Water Pollution* (1963) **7**, 787.
16. Akimoto, H., Finlayson, B. J., Pitts, Jr., J. N., unpublished data, 1971.
17. Cadle, R. D., Ledford, M., *Int. J. Air Water Pollution* (1966) **10**, 25.
18. Hales, J. M., Wilkes, J. O., York, J. L., *Atmos. Environ.* (1969) **3**, 657.

RECEIVED June 17, 1971.

Hydrogen Peroxide in the Urban Atmosphere

BRUCE W. GAY, JR. and JOSEPH J. BUFALINI

Environmental Protection Agency, Research Triangle Park, N. C. 27711

Hydrogen peroxide was measured in the atmospheres of Hoboken, N. J. and Riverside, Calif. At Hoboken, concentrations up to 4 pphm of hydrogen peroxide were determined in early afternoon hours during moderate photochemical smog formation. At Riverside, concentrations as high as 18 pphm were detected during severe smog formation. The concentration of hydrogen peroxide paralleled that of total oxidant.

In the presence of sunlight and oxides of nitrogen (NO_x), hydrocarbons react to form new products, some of which are called oxidants. The most commonly investigated photochemically produced oxidants in the urban atmosphere are ozone (O_3), nitrogen dioxide (NO_2), and peroxyacetylnitrate (PAN).

Recent laboratory studies show that another oxidant, peroxybenzoylnitrate (1), forms when systems containing aromatic hydrocarbons and NO_x are irradiated; however, this oxidant has not been found in the atmosphere. Other laboratory studies show that the photolyses of aldehydes (2, 3, 4) yield organic and inorganic hydroperoxides. Altshuller, Cohen, *et al.* (2) investigated the photolysis of propionaldehyde and found that ethyl hydroperoxide was a product. These workers also identified methyl hydroperoxide as a product in the photo-oxidation of acetaldehyde. Purcell and Cohen (4), studying the photo-oxidation products of formaldehyde (HCHO), found hydrogen peroxide (H_2O_2). Earlier work by Carruthers and Norrish (5) and later work by Horner and Style (6), concerning the photo-oxidation of HCHO, showed no H_2O_2. The products found in these studies were formic acid, CO, CO_2, and hydrogen; however, this work unlike the work of Purcell and Cohen was done at relatively high HCHO concentrations. Purcell and Cohen worked in the low concentration range of 1–30 ppm (v/v) whereas the others used millimeter partial pressures of HCHO.

Bufalini and Brubaker (7) studied the carbon fragments and oxidant formation resulting from the photo-oxidation of HCHO at two different wavelengths (3660 and 3130 A), with and without added NO_2 in the system. The only oxidant found when 12 ppm of HCHO in air was photo-oxidized at 3660 A was H_2O_2. When HCHO was photo-oxidized at 3660 A in the presence of NO_2, O_3 was also found. The maximum H_2O_2 concentration was that observed when no NO_2 was present. When HCHO was irradiated at 3130 A, 16 times more H_2O_2 was observed compared with irradiating at 3660 A. With added NO_2 and 3130 A radiation, only half as much H_2O_2 was found.

Aldehydes are primary and secondary pollutants in the urban atmosphere, and since laboratory studies indicate that they photo-oxidize to form peroxides, the presence of peroxides in the atmosphere seems evident.

In the early 1950's Haagen-Smit (8) stated that the oxidizing effect of smog resulted from the combined action of NO_2 and O_3 and peroxides. However no quantitative or qualitative study of hydroperoxides in the urban atmosphere has been reported.

Whether or not H_2O_2 is present in the urban atmosphere and, if it is, the concentration at which it exists are studied here.

Experimental

Hydrogen peroxide was determined in laboratory irradiated systems, in irradiated air samples containing auto exhaust collected at the entrance of the Lincoln Tunnel, and in ambient air samples at Hoboken, N. J., and

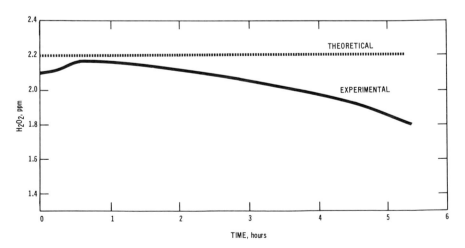

Figure 1.   Stability of hydrogen peroxide in a FEP Teflon bag

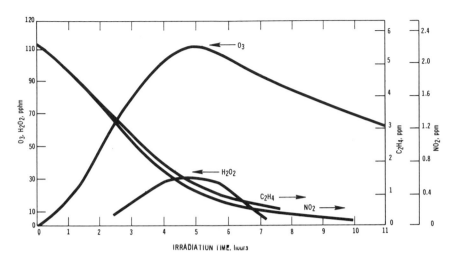

Figure 2.   Irradiation at 3660 A of ethylene with nitrogen dioxide

Riverside, Calif. The specific method developed by Cohen and Purcell (9) was used; here titanium IV and 8-quinolinol react with H_2O_2 to form a colored complex with absorption maximum at 450 mμ. There was no interference in the method for urban atmospheric levels of NO, NO_2, O_3, PAN, SO_2, and hydrocarbons.

Hydrogen peroxide concentrations in FEP Teflon bags were determined by the titanium IV–8-quinolinol and the catalyzed 1% potassium iodide colorimetric method (10).

At the New Jersey site O_3 was monitored with the Regener type instruments (11). In the laboratory it was determined with an instrument that measured the chemiluminescence from the ethylene–O_3 reaction (12).

Total oxidant readings in California were found on a Mast ozone instrument. In the low pphm concentration range of H_2O_2 the Mast instrument does not respond; in the ppm range of H_2O_2 the response is low and erratic.

Nitrogen dioxide was measured using the Saltzman method (13). Nitric oxide was oxidized to NO_2 and subsequently determined as NO_2.

Hydrocarbons were separated on a Porapack Q column and detected with a flame ionization detector.

All chemicals and gases were reagent grade and were used without purifying further.

Laboratory irradiations were conducted in a chamber fitted with 36 GE F-42-T6 blacklamps; the energy maximum was at 3660 A. Temperature in the chamber was maintained at 25 ± 2°C; Teflon FEP bags were used as reaction vessels in the chamber.

The light intensity of the chamber was measured as the rate of NO_2 photolysis in nitrogen (*14*). The first-order dissociation constant in the FEP Teflon bags was found to be 0.36 min^{-1}. The stability of H_2O_2 in FEP bags was investigated by injecting microliter quantities of a standardized 22.9% H_2O_2 solution into a metered stream of air while the bag was being filled. The theoretical concentration of peroxide was calculated from the amount of liquid injected and the volume of air used. The stability of H_2O_2 is shown in Figure 1. The concentrations observed experimentally were lower than the calculated theoretical concentration. This discrepancy was probably caused by destruction of H_2O_2 on the bag wall. The initial increase in H_2O_2 noted over the first half hour possibly resulted from desorption of H_2O_2 from the wall. When a bag is initially being filled, the surface–volume ratio is large, and some H_2O_2 apparently condenses on the wall. As the bag is filled, the surface–volume ratio decreases and H_2O_2 comes off the wall and into the gas phase.

Laboratory systems containing hydrocarbons and NO_x in air were irradiated and analyzed for oxidants. Four hydrocarbons that produced large amounts of HCHO per mole of reacted hydrocarbon were 1,3,5-trimethylbenzene, propylene, 1-butene, and ethylene. Hydrogen peroxide was detected in all four systems. Ozone was the major oxidant in these systems. Figure 2 shows the fate of a mixture of 5.5 ppm of ethylene (C_2H_4) and 2.2 ppm of NO_2 irradiated at 3660 A for 11 hours. The O_3

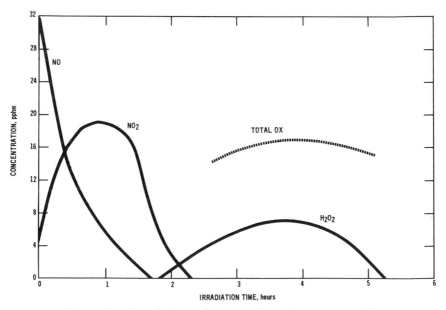

Figure 3.   Lincoln Tunnel air sample irradiated in sunlight

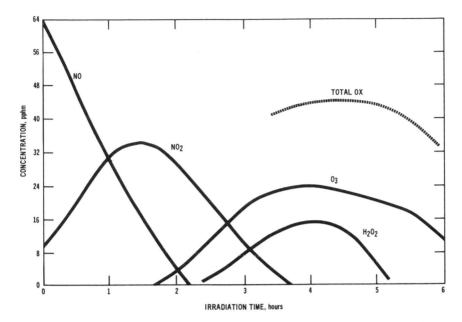

Figure 4.   Lincoln Tunnel air sample with nitric oxide, irradiated in sunlight

and H_2O_2 maximum concentrations occurred at almost the same time after 5 hours of irradiation.

During the summer of 1970, the Environmental Protection Agency had one of its mobile instrument trailers stationed in Hoboken, N. J. Air was collected in large plastic bags during the early morning hours at the time of heavy traffic at the New Jersey entrance to the Lincoln Tunnel. These bags of air were protected from sunlight and returned to the instrument trailer for irradiation and analysis. The irradiation was performed by exposing the bags to sunlight atop the trailer.

Before irradiation, one bag contained 6.0 ppm of CO, 4.1 ppm of methane, and 1.9 ppm of non-methane hydrocarbons; the results from irradiating this bag are shown in Figure 3. The total NO_x was 36 pphm, with 29.5 pphm as NO. After 1 hour of irradiation the NO_2 maximum was reached; after 2 hours all of the NO and most of the NO_2 had disappeared. Total oxidant maximum and H_2O_2 maximum were observed after 3¾ hours of irradiation.

To determine the effect of increasing the ratio of NO_x to hydrocarbon (NO_x/HC) on the formation of oxidant and rate of hydrocarbon reaction. more NO was injected into a sample of collected air like that used in Figure 3. The NO concentration was increased by a factor of 2; the results of irradiating the altered sample are shown in Figure 4. The NO_2 maximum was reached 30 minutes later than that for the unaltered air sample. After 3 hours of irradiation the system still contained NO_2. Most of the NO_x was lost from these systems during the irradiation with the nitrogen going to organic nitrates and nitric acid (15). Ozone and H_2O_2 maximums occurred after 4 hours of irradiation. The maximum H_2O_2 concentration in the altered system was twice H_2O_2 concentration in the unaltered system.

The ambient atmosphere at the mobile instrument site in Hoboken, N.J. contained up to 4 pphm of H_2O_2 on a day with high solar radiation and apparent photochemical smog formation. Hydrogen peroxide was observed between 12:00 A.M. and 2:00 P.M. On days when solar radiation was low because of cloud cover, no H_2O_2 was observed.

In August 1970 the urban atmosphere at Riverside, Calif. was sampled for H_2O_2 during days of photochemical smog formation. On the sixth, during a severe smog episode, concentrations of oxidant as high as 65 pphm were measured by the Mast ozone instrument (Figure 5). Hydrogen peroxide reached a maximum of 18 pphm during the episode at about the time the total oxidant was at its maximum.

Moving of the polluted air mass and changing of photochemical smog formation are seen in the changes in H_2O_2 concentrations at Riverside. Figure 6 shows the H_2O_2 concentration as a function of time of day at the Riverside site for three different days. On August 7 moderate to heavy smog buildup was observed for the air mass that had moved eastwardly over Riverside. Maximum total oxidant observed was about 30

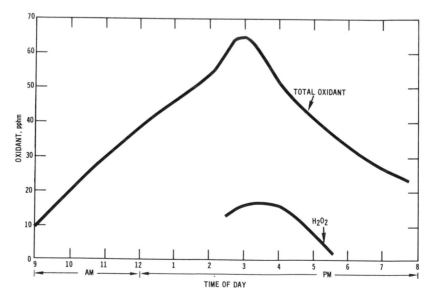

Figure 5.   Measured oxidant at Riverside, Calif. (August 6, 1970)

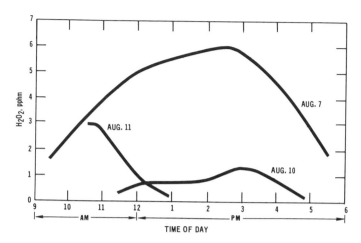

Figure 6.   Hydrogen peroxide concentrations at Riverside,
Calif. (August 1970)

pphm between 2:00 and 3:00 P.M. A maximum concentration of 6 pphm of H_2O_2 was found. The polluted air mass that formed west of Riverside on August 11 never reached the sampling site because of a wind direction change. Less photochemical smog on August 10 was indicated by increased visibility and lower oxidant readings; the maximum oxidant was observed about 3:00 P.M. when 1 pphm of H_2O_2 was observed.

Discussion

Analyses for H_2O_2 in laboratory-irradiated systems, in irradiated air samples including auto exhaust, and in the ambient polluted atmosphere have shown that H_2O_2 is present.

The formation of H_2O_2 is explained by a reaction sequence involving the photolysis of HCHO, nitrous acid, and NO_2. The photodissociation of HCHO is described by two processes (*16, 17, 18*):

$$HCHO \xrightarrow{h\nu} H + HCO \text{ and } HCHO \xrightarrow{h\nu} H_2 + CO$$

The first process is less important at longer wavelengths. The hydrogen atoms produced by it produce a hydroperoxy radical (Equation 1).

$$H + O_2 \xrightarrow{M} HO_2 \tag{1}$$

The hydroperoxy radical abstracts a labile hydrogen from a hydrocarbon or an aldehyde and forms H_2O_2.

$$HO_2\cdot + RCHO \rightarrow RCO + H_2O_2 \tag{2}$$

When CO, NO_x, and water vapor are present, hydrogen atoms form by the following sequence of reactions:

$$NO + NO_2 + H_2O \rightarrow 2\,HNO_2 \qquad (3)$$

$$\overset{h\nu}{HNO_2 \rightarrow OH + NO} \qquad (4)$$

$$\overset{h\nu}{NO_2 \rightarrow O + NO} \qquad (5)$$

$$O + RH \text{ or } RCHO \rightarrow OH + R \text{ or } RCO \qquad (6)$$

$$OH + CO \rightarrow CO_2 + H \qquad (7)$$

Hydrogen peroxide is then produced by Equations 1 and 2.

The H_2O_2 concentration curves in Figures 3 and 4 indicate that H_2O_2 does not form when NO is present. This is explained by the reaction

$$HO_2 + NO \rightarrow NO_2 + HO \qquad (8)$$

which is analogous to the oxidation of NO by RO_2, previously given as part of the overall reaction mechanism of photochemical air pollution (19).

The authors believe this is the first time H_2O_2 has been quantitatively and qualitatively identified in the urban atmosphere. Organic hydroperoxides probably are also formed in the urban atmosphere, but at lower concentrations.

Literature Cited

1. Heuss, J. M., Glasson, W. A., *Environ. Sci. Technol.* (1968) **2**, 1109.
2. Altshuller, A. P., Cohen, I. R., Purcell, T. C., *Can. J. Chem.* (1966) **44**, 2973.
3. Johnston, H. S., Heicklen, J. J., *J. Amer. Chem. Soc.* (1964) **86**, 4259.
4. Purcell, T. C., Cohen, I. R., *Environ. Sci. Technol.* (1967) **1**, 845.
5. Carruthers, J. E., Norrish, R. G. W., *J. Chem. Soc.* (1936) 1036.
6. Horner, E. C. A., Style, D. W. G., *Trans. Faraday Soc.* (1954) **50**, 1197.
7. Bufalini, J. J., Brubaker, K., *Symp. Chem. Reactions Urban Atmospheres*, Research Laboratories, General Motors Corp., Warren (Oct. 6–7, 1969).
8. Haagen-Smit, A. J., *Ind. Eng. Chem.* (1952) **44**, 1342.
9. Cohen, I. R., Purcell, T. C., *Anal. Chem.* (1967) **39**, 131.
10. Cohen, I. R., Purcell, T. C., Altshuller, A. P., *Environ. Sci. Technol.* (1967) **1**, 247.
11. Regener, V. H., *J. Geophys. Res.* (1960) **65**, 3975.
12. Nederbraght, G. W., *Nature* (1965) **206**, 87.
13. Saltzman, B. E., *Anal. Chem.* (1954) **26**, 1949.
14. Tuesday, C. S., "Chemical Reactions in the Lower and Upper Atmosphere," p. 1–49, R. D. Cadle, Ed., Interscience, New York, 1961.
15. Gay, Jr., B. W., Bufalini, J. J., *Environ. Sci. Technol.* (1971) **5**, 422.

16. Calvert, J. G., Pitts, Jr., J. N., "Photochemistry," p. 371, Wiley, New York, 1966.
17. Degraff, B. A., Calvert, J. G., *J. Amer. Chem. Soc.* (1967) **89**, 2247.
18. McQuigg, R. D., Calvert, J. G., *J. Amer. Chem. Soc.* (1969) **91**, 1590.
19. Leighton, P. A., "Physical Chemistry," Vol. IX, p. 218, Academic, New York, 1961.

RECEIVED July 26, 1971. Mention of product or company names does not constitute indorsement by the Environmental Protection Agency.

12

The Role of Ozone in the Photooxidation of Propylene in the Presence of NO₂ and O₂

S. JAFFE and R. LOUDON

California State College, Los Angeles, Calif. 90032

Mixtures of propylene and NO₂ were photolyzed at 3660A with varying amounts of O₂, ranging from zero to one atmosphere. The quantum yields for NO₂ consumption were measured along with several of the reaction products. The changes in Φ(NO)₂ and product production rates as a function of O₂ are attributed to the formation of O₃ and the various reactions of O₂ with the intermediates and free radicals. A mechanism is presented to account for the observed behavior.

The effect of molecular oxygen on the photolysis rates of NO₂ and the reaction of ozone with propylene have been studied separately; also the photolysis of NO₂ in the presence of propylene and absence of O₂ have been studied. This work tries to determine the effect of a systematic variation of molecular oxygen on the photooxidation of propylene in the presence of NO₂. This process should be analogous to that which occurs after NO₂ concentration is maximized when NO and olefins are irradiated in air.

The photolysis of mixtures of NO, NO₂, and O₂ by Ford (1) shows that the quantum yields for NO₂ decomposition are inversely proportional to the O₂ pressure, and the effect is mainly attributed to the formation of O₃ when O atoms react with O₂.

In other studies the effect of varying O₂ was observed by Sato and Cvetanovic (2) in oxidizing *cis*-2-pentene. They produced O atoms by the mercury sensitized dissociation of N₂O and by photolyzing NO₂. They showed that the fraction of NO₂ consumed was decreased as the O₂ pressure increased and that the production rates of products were affected by the O₂ pressure. The production rates of epoxides decreased, and the production rates of aldehydes increased as the O_2 pressure increased.

Wei and Cvetanovic (3) measured the reaction rates of ozone with olefins in the presence and absence of molecular oxygen and found that the relative rate constants were lower when O_2 was absent. They also found that the reactions showed a 1:1 stoichiometry in the absence of O_2 and 1:1.4 to 1:2.0 stoichiometry for different olefins in the presence of O_2.

Several investigators (4) have studied the photooxidation of olefins in air at atmospheric pressure—*e.g.*, Altshuller (5) and co-workers investigated the propylene–NO system.

An attempt is made to correlate previous observations and ours for the oxidation of propylene in the absence of O_2 (6) with the present work on the oxidation of propylene as a function of O_2.

The processes occurring in each study should occur simultaneously in the propylene, NO_2, O_2 system. Varying O_2 should show the competition among NO_2, C_3H_6, and O_2 for O atoms and also the competition between O_3 and O to oxidize C_3H_6.

Experimental

The experimental procedure was similar to that reported earlier (6, 7). Samples of NO_2, C_3H_6, and O_2 were introduced into a 4 foot long, 2 inch diameter quartz tube by standard high vacuum techniques at 24 ± 1°C. The cell was irradiated with a Hanovia S-100 mercury arc. The light was filtered by a series of interference and blocking filters to allow only a narrow band near 3660A to pass through the cell. The transmitted light struck a photomultiplier tube whose output was measured with a Fluke model 885 A dc differential voltmeter. The increase in voltage as NO_2 dissociated was recorded on a Moseley Model 680 potentiometer. The voltage-time curves were linear, so that the slopes of the curves could be correlated with the NO_2 dissociation rate. The voltage measured with known concentrations of pure NO_2 served to calibrate the NO_2 concentration and allowed the light intensity to be calculated from known values (8) of the quantum yields for the dissociation of pure NO_2.

After irradiating, the gases were pumped out of the cell through a series of traps at $-115°C$ and $-195°C$. By this means N_2 and O_2 were separated from the products, and two separate groups of products were trapped for subsequent chromatographic analysis. The trap containing the excess NO_2 was treated with mercury to remove the NO_2 before injecting that sample into the chromatograph.

The principal method of analyzing was by a Loenco Model 15 B gas chromatograph fitted with thermal conductivity and flame ionization detectors. The analysis was made on Poropak Q, 80–100 mesh, packed in a 1/8 inch diameter, 12 foot long stainless steel column. Calibrating and identifying the products was accomplished by analyzing known amounts of authentic CP reagents. Integration of the curves was done by a disc integrator.

The O_2 and N_2 used here were specially prepared by the Air Reduction Co. and placed in cleaned gas cylinders. They were further purified by passing the gases through a train containing Ascarite to remove CO_2, followed by $Mg(ClO_4)_2$ to remove H_2O, and then by Drierite. All other materials were CP and were further purified by trap to trap distillation. NO_2 was stored at $-195°C$ and redistilled after being treated with O_2 whenever a slight blue color indicated the presence of NO as N_2O_3.

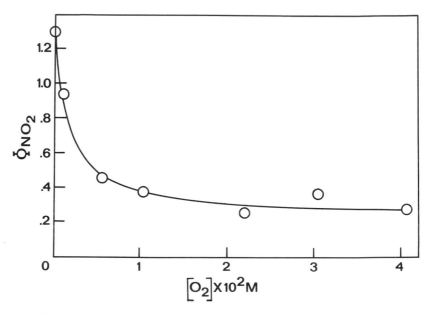

Figure 1.   Quantum yields as a function of oxygen concentration

Results

A series of reactions was carried out at $24 \pm 1°C$ with an average concentration of NO_2 equal to 5.40×10^{-5} M and an average concentration of C_3H_6 equal to 2.03×10^{-4} M. The oxygen concentrations varied from 0 to 4.09×10^{-2} M (1 atm). The total pressure was kept constant at one atmosphere by adding N_2. This set of experiments was irradiated for 3000 seconds. Quantum yields for NO_2 consumption were determined and are shown in Figure 1; they decrease sharply as O_2 is added and approach a constant value after about 1×10^{-2} moles per liter of O_2 have been added. These results agree with those of Ford (*1*) on how O_2 affects pure NO_2 samples and indicate that competitive reactions of O atoms with O_2 and C_3H_6 are taking place and decrease the reaction rate of O atoms with NO_2 as the O_2 concentration increased. The subsequent reactions of NO_2 with several free radicals are also inhibited by the reaction of O_2 with the free radicals where possible and further reduce the quantum yield. NO_2 reforming reactions with O_3 and NO_3 plus NO also have this effect.

The products of the reaction included NO, CO_2, H_2O, CH_3CHO, CH_3COCH_3, CH_3CH_2CHO, $CH_3\overset{\displaystyle O}{\overset{\displaystyle \triangle}{C}}HCH_2$, CH_3NO_2, CH_3ONO_2, CH_3CH_2-NO_2, $CH_3CH_2ONO_2$, 2-$C_3H_7NO_2$, and minor quantities of higher molecular weight species. The presence of CO was determined mass spectrometrically, but its production rate was not measured. There must also have been appreciable amounts of CH_2O present, but it polymerized to paraformaldehyde before the chromatographic column. A white solid in the cold traps appeared, resulting from the CH_2O. The CH_2O along with CH_3CHO should have been the major products. Altshuller *et al.* (5) found equal quantities of CH_2O and CH_3CHO when they studied the NO–propylene–oxygen system, and our results based on the mechanism in which CH_3CHO and CH_2O are similarly produced are assumed to be similar. The concentration of CH_3CHO (*see* Figure 2) was an order of magnitude greater than most other products, confirming this assumption.

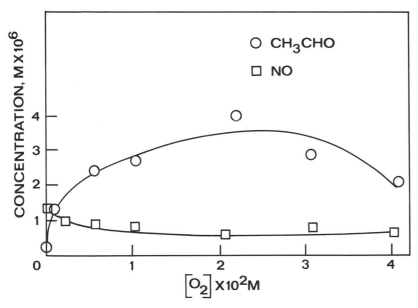

Figure 2. NO and CH₃CHO as a function of oxygen concentration

O_2 affected the production of the products as shown in Figures 2 through 5 for a representative set of experiments. They are also represented in Table I by least squares equations of the form,

Table I.   Least Squares Relations for

Product P	a	b
NO	8.66×10^{-7}	3.09×10^{-5}
CO_2	-1.19×10^{-9}	4.97×10^{-6}
CH_3CHO	2.34×10^{-7}	1.22×10^{-4}
$CH_3\overset{\overset{O}{\|}}{C}CH_3$	6.71×10^{-8}	2.27×10^{-5}
$CH_3\overset{O}{\overset{\triangle}{CH}CH_2}$	1.66×10^{-7}	5.11×10^{-5}
C_2H_5CHO	2.47×10^{-7}	4.16×10^{-5}
CH_3ONO_2	2.25×10^{-7}	2.34×10^{-5}
CH_3NO_2	1.68×10^{-7}	-1.37×10^{-5}
$C_2H_5ONO_2$	5.13×10^{-8}	1.46×10^{-5}

$$P = a + b[O_2] + C[O_2]^2 + d[O_2]^3 + e[O_2]^4$$

The trends in product concentrations as a function of O_2 are reviewed below.

The decrease in NO as O_2 increases (Figure 2) agrees with the quantum yield data and also shows NO rapidly reacting with O_3. The increase in CO_2 (Figure 3) probably shows free radicals reacting with O_2. The relatively large, increasing concentration of CH_3CHO (Figure 2)

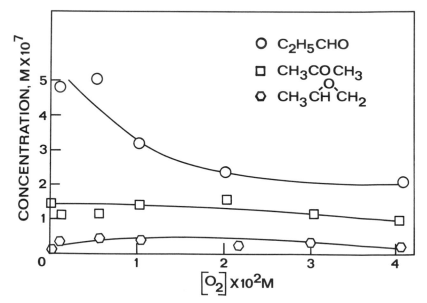

Figure 3. CH_3ONO_2 and CO_2 as a function of oxygen concentration

Producing Products as a Function of [O_2]

c	d	e
-4.35×10^{-3}	1.33×10^{-1}	-1.16
8.88×10^{-4}	-3.40×10^{-2}	2.98×10^{-1}
-3.91×10^{-3}	-7.54×10^{-2}	3.05
-1.88×10^{-3}	5.75×10^{-2}	-6.04×10^{-1}
-4.48×10^{-3}	1.45×10^{-1}	-1.66
-5.00×10^{-3}	1.89×10^{-1}	-2.27
2.16×10^{-3}	-2.40×10^{-1}	4.34
3.66×10^{-4}	4.49×10^{-3}	-1.83×10^{-1}
-9.68×10^{-4}	2.58×10^{-2}	-2.65×10^{-1}

supports the increasing importance of O_3–olefin reactions and the subsequent reactions of zwitterions which produce CH_3CHO and CH_2O as principal products. As the products resulting from O_3 reactions increase, those resulting from the oxidation of C_3H_6 by O atoms, which produce CH_3COCH_3, $CH_3\overset{O}{\overset{\triangle}{C}HCH_2}$, and C_2H_5CHO (Figure 4), decreases as O_2 increases.

The concentration of CH_3ONO_2 (Figure 3) is relatively high and increases slightly with increasing O_2. This may show the increase in the availability of CH_3 radicals in this process. However CH_3 radicals are converted to CH_3O radicals more rapidly than they recombine with NO_2 so that the production of CH_3NO_2 (Figure 5) decreases. This might indicate another unexpected path for CH_3O forming in the presence of O_2. The production of $C_2H_5ONO_2$ (Figure 5) is not similarly affected since C_2H_5 radicals are assumed to be produced from the dissociation of excited propionaldehyde, the supply of which would be decreased as O_2 increased.

The production of $C_2H_5NO_2$ and $CH_3CH(NO_3)CH_3$ were not quantitatively determined, but their presence in chromatograms was confirmed. They were not greatly affected by O_2 changes and are not important for short irradiation times.

Discussion

This system's behavior is represented by a mechanism that combines the reactions associated with irradiating NO_2 and C_3H_6 and with the mechanism for oxidizing C_3H_6 by O_3. The process also includes the reactions of O_2 and NO_2 with all of the free radicals and intermediates with

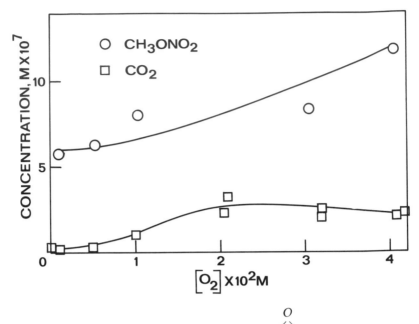

Figure 4. C_2H_5CHO, CH_3COCH_3, and $CH_3\overset{O}{\overset{/\backslash}{CH}}CH_2$ as a function of oxygen concentration

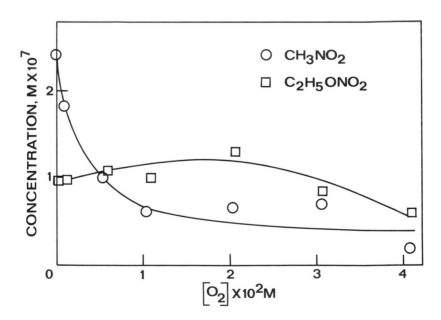

Figure 5. CH_3NO_2 and $C_2H_5ONO_2$ as a function of oxygen concentration

which they interact. This represents the process in its early stages so that the many secondary reactions need not be treated. Thus free radicals and intermediates only react with the major components, NO_2, C_3H_6, and O_2, and not with each other. We give the following mechanism along with the estimated heats of reaction to support the feasibility of the various steps in this study. The data for estimating the heats of reaction were taken from References 9, 10, and 11.

$$\Delta H°, Kcal$$

		$\Delta H°$,Kcal	
$NO_2 + h\nu$	$\rightarrow NO + O$		(Ia)
$O + NO_2$	$\rightarrow NO + O_2$	-46	(1)
$O + NO_2 + M$	$\rightarrow NO_3 + M$	-50	(2)
$NO_3 + NO$	$\rightarrow 2 NO_2$	-22	(3)

$$O + C_3H_6 \rightarrow CH_3\overset{\cdot}{C}H\overset{\overset{\displaystyle \cdot O}{|}}{C}H_2 \qquad -39.5 \qquad (4)$$

$$\rightarrow CH_3\overset{\overset{\displaystyle O\cdot}{|}}{C}H\overset{\cdot}{C}H_2 \qquad -41.5 \qquad (5)$$

$O + O_2 + M$	$\rightarrow O_3 + M$	-25.6	(6)
$O_3 + NO_2$	$\rightarrow NO_3 + O_2$	-25.4	(7)
$O_3 + NO$	$\rightarrow NO_2 + O_2$	-47.6	(8)
$O_3 + C_3H_6$	$\rightarrow CH_3\overset{+}{C}HOO^- + CH_2O$	-33	(9)
	$\rightarrow H\overset{+}{C}HOO^- + CH_3CHO$	-40	(10)
$CH_3\overset{+}{C}HOO^- + NO_2$	$\rightarrow NO_3 + CH_3CHO$	-64	(11)
$H\overset{+}{C}HOO^- + NO_2$	$\rightarrow NO_3 + CH_2O$	-57	(12)

$$C_3H_6 + CH_3\overset{+}{C}HOO^- \rightarrow CH_3\overset{\cdot}{C}H\overset{\overset{\displaystyle \cdot O}{|}}{C}H_2 + CH_3CHO \qquad -52.6 \qquad (13)$$

$$\rightarrow CH_3\overset{\overset{\displaystyle O\cdot}{|}}{C}H\overset{\cdot}{C}H_2 + CH_3CHO \qquad -54.6 \qquad (14)$$

$$H\overset{+}{C}HOO^- + C_3H_6 \rightarrow CH_3\overset{\cdot}{C}H\overset{\overset{\displaystyle \cdot O}{|}}{C}H_2 + CH_2O \qquad -45.6 \qquad (15)$$

$$\rightarrow CH_3\overset{\overset{\displaystyle O\cdot}{|}}{C}H\overset{\cdot}{C}H_2 + CH_2O \qquad -47.6 \qquad (16)$$

$\triangle H°,\text{Kcal}$

$$CH_3\overset{+}{C}HOO^- + O_2 \quad \rightarrow CH_3CHO + O_3 \qquad\qquad -38.7 \qquad (17)$$

$$H\overset{+}{C}HOO^- + O_2 \quad \rightarrow CH_2O + O_3 \qquad\qquad -31.7 \qquad (18)$$

$$CH_3\overset{O\,\cdot}{\underset{\cdot}{C}}HCH_2 \quad \underset{\longleftarrow}{\rightarrow} CH_3\overset{O}{\overset{\triangle}{C}H-CH_2}{}^* \qquad\qquad -58 \qquad (19)$$

$$\hookrightarrow CH_3\overset{O}{\overset{\|}{C}}CH_3{}^* \qquad\qquad -74.7 \qquad (20)$$

$$CH_3\overset{\cdot\,O}{\underset{\cdot}{C}}HCH_2 \quad \underset{\longleftarrow}{\rightarrow} CH_3\overset{O}{\overset{\triangle}{C}HCH_2}{}^* \qquad\qquad -60 \qquad (21)$$

$$\hookrightarrow CH_3CH_2CHO^* \qquad\qquad -78 \qquad (22)$$

$$CH_3\overset{O}{\overset{\triangle}{C}HCH_2}{}^* + M \quad \rightarrow CH_3\overset{O}{\overset{\triangle}{C}HCH_2} + M \qquad\qquad (23)$$

$$CH_3\overset{O}{\overset{\|}{C}}CH_3{}^* + M \quad \rightarrow CH_3\overset{O}{\overset{\|}{C}}CH_3 + M \qquad\qquad (24)$$

$$CH_3\overset{O}{\overset{\|}{C}}CH_3{}^* \quad \rightarrow \cdot CH_3 + \cdot \overset{O}{\overset{\|}{C}}CH_3 \qquad\qquad +78.4 \qquad (25)$$

$$CH_3CH_2CHO^* + M \quad \rightarrow CH_3CH_2CHO + M \qquad\qquad (26)$$

$$CH_3CH_2CHO^* \quad \rightarrow CH_3\overset{\cdot}{C}H_2 + \overset{\cdot}{C}HO \qquad\qquad +87.2 \qquad (27)$$

$$CH_3 + O_2 \quad \rightarrow CH_2O + OH \qquad\qquad -50.4 \qquad (28)$$

$$CHO + O_2 \quad \rightarrow CO_2 + OH \qquad\qquad -92.7 \qquad (29)$$

$$CH_3CO + O_2 \quad \rightarrow CH_3\overset{O}{\overset{\|}{C}}O_2 \qquad\qquad (4) \qquad (30)$$

$$CH_3\overset{O}{\overset{\|}{C}}O_2 + O_2 \quad \rightarrow CH_3\overset{O}{\overset{\|}{C}}O \cdot + O_3 \qquad\qquad -11 \qquad (31)$$

$$CH_3\overset{O}{\overset{\|}{C}}-O \quad \rightarrow CH_3 + CO_2 \qquad\qquad -17 \qquad (32)$$

$$CH_3\overset{\cdot}{C}H_2 + O_2 \quad \rightarrow C_2H_5O_2 \qquad\qquad -24.2 \qquad (33)$$

$$NO_2 + CH_3 \quad \overset{\ulcorner}{\rightarrow} CH_3NO_2 \qquad\qquad -57.9 \qquad (34)$$

$$\text{L}{\rightarrow} \text{CH}_3\text{O} + \text{NO} \qquad\qquad -18.9 \qquad (35)$$

$$\text{NO}_2 + \text{C}_2\text{H}_5 \quad \rightarrow \text{C}_2\text{H}_5\text{NO}_2 \qquad\qquad -58 \qquad (36)$$

$$\text{L}{\rightarrow} \text{C}_2\text{H}_5\text{O} + \text{NO} \qquad\qquad -21.1 \qquad (37)$$

$$\text{NO}_2 + \text{CH}_3\text{O} \quad \rightarrow \text{CH}_3\text{ONO}_2 \qquad\qquad -36.1 \qquad (38)$$

$$\text{NO}_2 + \text{C}_2\text{H}_5\text{O} \quad \rightarrow \text{C}_2\text{H}_5\text{ONO}_2 \qquad\qquad -36.3 \qquad (39)$$

$$\text{NO}_2 + \text{CH}_3\text{CO} \quad \rightarrow \text{CH}_3\text{CO}_2 + \text{NO} \qquad -27.1 \qquad (40)$$

$$\text{NO}_2 + \text{CHO} \quad \rightarrow \text{CO} + \text{HNO}_2 \qquad\qquad -60 \qquad (41)$$

$$\text{L}{\rightarrow} \text{NO} + \text{HCO}_2 \qquad\qquad -4 \qquad (42)$$

$$\text{NO}_2 + \text{HCO}_2 \quad \rightarrow \text{CO}_2 + \text{HNO}_2 \qquad\qquad -110 \qquad (43)$$

The oxidation process is initiated by the photo-dissociation of NO_2 yielding O (^3P) atoms which enter into a series of reactions discussed by Cvetanovic (*12*) and applied to the photooxidation of propylene (*6*). Equations 4, 5, and 19 through 27 consist of adding oxygen to the double bond in propylene, forming a diradical, and the subsequent rearranging of the diradical to yield propionaldehyde, propionoxide, acetone, and the several free radicals. The free radicals react with NO_2 in Equations 34 through 43 as described earlier (*6*). The reactions of several of the free radicals with O_2 have been reviewed by Heicklen (*13*), and those reactions that seem feasible here are indicated in Equations 28 through 33.

Free radicals may also react with olefins (*13*), in general,

$$\Delta H°, \text{Kcal}$$

$$\text{RO} + \text{C}_3\text{H}_6 \qquad \rightarrow \text{RO} \cdot \text{C}_3\text{H}_6 \qquad\qquad -13 \qquad (44)$$

$$\text{RO}_2 + \text{C}_3\text{H}_6 \qquad \rightarrow \text{RO}_2 \cdot \text{C}_3\text{H}_6 \qquad\qquad -12 \qquad (45)$$

The reactions of olefins with OH are discussed by Leighton (*14*) and consist of:

$$\text{OH} + \text{C}_3\text{H}_6 \qquad \rightarrow \text{HO} \cdot \text{C}_3\text{H}_6 \qquad\qquad (46)$$

$$\rightarrow \text{H}_2\text{O} + \overset{\bullet}{\text{C}_3\text{H}_5} \qquad\qquad (47)$$

Although Equations 44, 45, and 46 are feasible, they cannot be proved by the products found and are not included in the main mechanism. The abstraction of hydrogen by OH must be more prevalent than indicated by Equation 47 alone since water was found in relatively large amounts. Abstraction of H from CH_3CHO and CH_2O by

$$\text{RCHO} + \text{OH} \rightarrow \text{H}_2\text{O} + \text{RCO} \qquad\qquad (48)$$

may be important.

Any of the changes in product production as a function of O_2 principally results in the increasing importance of Equation 6. A simple calculation, based on a steady state approximation for O atoms and for O_3 in this mechanism, indicates that the rate of Equation 1 is only about 1.6 times faster than Equation 6 when O_2 approaches 4×10^{-2} molar and the rate constants in Table II are used. Therefore ozone should be formed at a sufficient rate under the present conditions to contribute to the mechanism as shown.

Table II.   Reaction Rate Constants at 25°C

Reaction Number	Rate Constant	Reference
1	3.3×10^9 liter m^{-1}sec^{-1}	24
2	1.0×10^{11} liter m^{-2}sec^{-1}	25
3	5.6×10^9 liter m^{-1}sec^{-1}	26
4,5	1.7×10^9 liter m^{-1}sec^{-1}	12,24 [a]
6	6.9×10^7 liter^{2}m^{-2}sec^{-1}	27
7	4.3×10^4 liter m^{-1}sec^{-1}	28
8	2.8×10^7 liter m^{-1}sec^{-1}	1
9,10	5.1×10^3 liter m^{-1}sec^{-1}	16
23,24,26	1×10^{11} liter m^{-1}sec^{-1}	6
25	2×10^8 sec^{-1}	6
27	1×10^8 sec^{-1}	6
28	6×10^7 liter m^{-1} sec^{-1}	13
33	6.3×10^8 liter m^{-1}sec^{-1}	13
34	1.7×10^9 liter m^{-1}sec^{-1}	27
35	3.3×10^9 liter m^{-1}sec^{-1}	27

[a] Calculated from results in 12 with new value for equation 1.

The reactions of O_3 with C_3H_6, 9 and 10, were based on a review of the mechanism of ozonolysis by Murray (15). The accepted mechanism in the gas phase is similar to that commonly referred to as the Criegee mechanism in the liquid phase. The attack of O_3 on C_3H_6 is electrophilic, and Vrbaski and Cvetanovic (16) have correlated the electrophilic behavior with that of oxygen atom–olefin reactions. Both reaction rates correlate with the ionization potentials of a series of olefins.

The preferred structure of the initial adduct is the one that results from a one-step, cis-addition to the double bond,

This structure is preferred by several investigators (17, 18, 19, 20, 21).

Although there is no evidence for the zwitterion in the gas phase, evidence in the liquid phase for $\overset{+}{C}$—O—O$^-$ or for $> = \overset{+}{O}$—O$^-$ has been advanced by Criegee (22).

Writing the intermediates as zwitterions in this case is a formalism since other forms have been used. Benson (23) gives the sequence as follows:

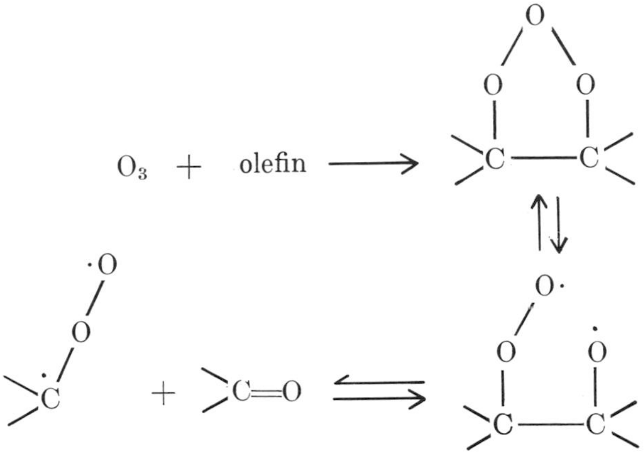

In the gas phase formation of ozonide would be unlikely, and the intermediates should live long enough to react with NO_2 as in Equations 11 and 12, with C_3H_6 as in 13, 14, 15, and 16, and with O_2 as in 17 and 18. However preliminary calculations using Equation 51 indicate that reactions 13, 14, 15, and 16 are too slow to be important in this process. Reactions of intermediates, as in 17 and 18, explain the non-stoichiometric ratio of olefin to O_3 in the presence of O_2 as observed by Cvetanovic (3) and others. Ozone may also be formed upon reacting with peroxyacyl radicals such as in Equation 31.

Also the intermediates may decompose unimolecularly:

$$CH_3\overset{+}{C}HOO^- \quad \begin{array}{l} \longrightarrow CH_4 + CO_2 \qquad (49) \\ \longrightarrow CH_3OH + CO \qquad (50) \end{array}$$

but these reactions make minor contributions to the mechanism.

In the present system ozone should not be able to build up in concentration as usual in systems at low NO_2 concentrations. Reaction with NO_2 as in Equation 7 and with NO as in Equation 8 should tend to keep the ozone concentration at the steady state level.

NO_2 is rapidly regenerated as in 3 and 8 so that along with the decrease in the rate of Equations 1 and 2 because of the competition for O atoms by O_2, the decrease in quantum yield (*see* Figure 1) is expected as the O_2 concentration increases.

When the steady state approximation is applied to free radicals and intermediates in Equations 1_a and 1 through 43, a complex rate equation is obtained. If $k_4 = k_5$, $k_9 = k_{10}$, $k_{11} = k_{12}$, $k_{13} = k_{14} = k_{15} = k_{16}$, $k_{17} = k_{18}$, $k_{19} = k_{20} = k_{21} = k_{22}$, $k_{23} = k_{24} = k_{26}$, $k_{25} = k_{27}$, $k_{34} = k_{36}$, $k_{35} = k_{37}$, and $k_{38} = k_{39}$ is assumed, Equation 51 results.

$$\frac{-d(NO_2)}{dt} = I_a + A - [B + T](O_3) - \frac{C(O_3)}{X}$$

$$+ D\left[Y + \frac{E(O_3)}{X}\right]\left[\frac{10F}{Z} + \frac{5F}{W} + \frac{G}{V} + \frac{H}{U}\right] \qquad (51)$$

The terms in Equation 51 are:

$$A = [k_1 - k_2 (M)] (NO_2)(O)$$
$$B = k_7(NO_2)$$
$$C = 2\,k_9k_{11}(C_3H_6)(NO_2)$$
$$D = \frac{1}{2[k_{24}(M) + k_{25}]}$$
$$E = 2\,k_9k_{13}(C_3H_6)$$
$$F = k_{25}k_{34}(NO_2)$$
$$G = k_{25}k_{40}(NO_2)$$
$$H = [k_{41} + 2\,k_{42}](NO_2)$$
$$X = [k_{11}(NO_2) + k_{13}(C_3H_6) + k_{17}(O_2)]$$
$$Y = k_4(O)(C_3H_6)$$
$$Z = [k_{28}(O_2) + 3\,k_{34}(NO_2)]$$
$$T = k_8(NO)$$
$$U = [k_{29}(O_2) + (k_{41} + k_{42})(NO_2)]$$
$$V = [k_{30}(O_2) + k_{40}(NO_2)]$$
$$W = [k_{33}(O_2) + 2k_{34}(NO_2)]$$
$$(O) = \frac{I_a}{[k_1(NO_2) + k_2(NO_2)(M) + (k_4 + k_5)(C_3H_6) + k_6(O_2)(M)]}$$

Preliminary calculations show that this equation explains the qualitative trend in the quantum yield data (Figure 1) and that it shows the dependence on O_2 and O_3. The steady state concentration of O atoms is inversely proportional to the O_2 concentration because of Equation 6, varying from 1.3 to 1.2×10^{-15} moles liter^{-1}. The negative terms in (O_3) become more important as the O_3 concentration increases, as it does when O_2 increases. This effectively reduces the rate of loss of NO_2. The last set of terms are all inversely proportional to O_2 so that they all decrease as O_2 increases. Since they are multiplied by $(Y + E(O_3)/X)$, the net effect is that the product of this factor and the last set of terms remains almost constant. A net decrease in the loss rate of (NO_2) occurs as O_2 is increased. Here the value of I_a, the rate of absorption of light, was almost constant at an average of 1.43×10^{-9} Einsteins per second. When it is factored out of the equation and divided into $-d(NO_2)/dt$, the quantum yield, which shows the same trends as described above, is obtained.

A similar treatment is being applied to all of the production rates of the products listed in Table I. We derived equations like 51 for each of the products and are now fitting the data in Figures 1 through 5 into them. The equations in Table I may be used to obtain $d(P)/d(O_2)$ for each product, and these rates may be inserted into the rate equations that have been derived from the mechanism. It is hoped that some of the rate constants that are missing from Table II may be evaluated in this manner. Results of this endeavor will soon be reported.

Acknowledgment

The authors are grateful for the support of the National Air Pollution Control Administration under grant number 5 R01 APOO 462. Also we wish to thank Eddy Wan for his assistance.

Literature Cited

1. Ford, H. W., Endow, N., *J. Chem. Phys.* (1957) **27**, 1156.
2. Sato, S., Cvetanovic, R. J., *Can. J. Chem.* (1959) **37**, 953.
3. Wei, Y. K., Cvetanovic, R. J., *Can. J. Chem.* (1963) **41**, 913.
4. Altshuller, A. P., Bufalini, J. J., *Photochem. Photobiol.* (1965) **4**, 97.
5. Altshuller, A. P., Kopczynski, S. K., Louneman, W. A., Becker, T. L., Slater, R., *Environ. Sci. Technol.* (1967) **1**, 889.
6. Jaffe, S., Grant, R. C. S., *J. Chem. Phys.* (1969) **50**, 3477.
7. Jaffe, S., Keith, J., *J. Chem. Phys.* (1968) **48**, 2805.
8. Ford, H. W., Jaffe, S., *J. Chem. Phys.* (1963) **38**, 2935.
9. Calvert, J. G., Pitts, J. N., Jr., "Photochemistry," pp. 815–826, Wiley, 1967.
10. Kerr, J. A., *Chem. Rev.* (1966) **66**, 465.
11. Benson, S. W., Cruikshank, F. R., Golden, D. M., Hauger, G. R., O'Neal, H. E., Rodgers, A. S., Shaw, R., Walsh, R., *Chem. Rev.* (1968) **69**, 279.
12. Cvetanovic, R. J., *Advan. Photochem.* (1963) **1**, 115–182.
13. Heicklen, J., Proc. Intern. Oxidation Symp., p. 343, Vol. I, San Francisco (Aug. 28–Sept. 1, 1967).
14. Leighton, P. A., "Photochemistry of Air Pollution," pp. 226, 227, Academic, New York, 1961.
15. Murray, R. W., *Accounts of Chem. Res.* (1968) **1**, 313.
16. Vrbaski, T., Cvetanovic, R. J., *Can. J. Chem.* (1960) **38**, 1053.
17. Criegee, R., Schröder, G., *Chem. Ber.* (1960) **93**, 689.
18. Greenwood, F. L., *J. Org. Chem.* (1964) **29**, 1321.
19. Greenwood, F. L., *J. Org. Chem.* (1965) **30**, 3108.
20. DeMore, W. B., *Int. J. Chem. Kinetics* (1969) **1**, 209.
21. Thompson, J. A., Shoulder, B. A., *J. Amer. Chem. Soc.* (1966) **88**, 4098.
22. Criegee, R., "Peroxide Reaction Mechanisms," p. 29, J. O. Edwards, Ed., Interscience, New York, 1962.
23. Benson, S. W., *Advan. Chem. Ser.* (1968) **77**, 74.
24. Klein, S. F., Herron, J. T., *J. Chem. Phys.* (1964) **41**, 1285.
25. Hall, T. C., Jr., Blout, F. E., *J. Chem. Phys.* (1952) **20**, 1745.
26. Schott, G., Davidson, N., *J. Amer. Chem. Soc.* (1958) **80**, 1841.
27. Benson, S. W., Axworthy, A. E., *J. Chem. Phys.* (1957) **26**, 1718.
28. Johnston, H. S., Yost, P. M., *J. Chem. Phys.* (1949) **17**, 366.
29. Phillips, L., Shaw, R., *Symp. Combust. 10th*, University of Cambridge, 453–461 (1965).

RECEIVED May 10, 1971.

INDEX

INDEX